I0479296

Fluid Kinematics

With

Bernoulli's Equation

FUNDAMENTALS AND APPLICATIONS

PAPRI BHATTACHARJEE

ISBN: 9798387000850

ACKNOWLEDGEMENTS

❖ **Writing a book is harder than I thought and more rewarding than I have ever imagined.** At this moment of accomplishment, I am greatly indebted to Amazon KDP University, and Mr. S. Bathla for providing me all the help. This book would not have been possible without their support and involvement.

❖ My earnest thanks to Prof. Y.A. Cengel, Prof. J. M. Cimbala, Prof. R.K. Bansal, Prof. M.D. Raisinghania, and last but not least Prof. S. K. Sharma for making me capable of writing this book. Without their guidance, it was almost impossible to write this book.

❖ Finally, I acknowledge the people who mean a lot to me, my parents, for showing faith in me and for their selfless love, care, pain, and sacrifice. I thank the Almighty for giving me the strength and patience to work on this book so that today I stand confidently.

TABLE OF CONTENTS

1. INTRODUCTION

DEFINITION OF BERNOULLI PRINCIPLE

Have you wondered how the water flows from the higher region to lower region in a mountain or in waterfall? This happen because the fluid moves from the high-pressure zone to low pressure zone. The fluid elements experiences more energy which results in accelerate their motion towards the low-pressure area. This phenomenon is based on the Bernoulli's principle.

The Bernoulli's equation is the mathematical relation between pressure, velocity, and elevation which derivable based on following assumptions. Due to its simplicity, it is widely used in many real-life applications which we will discuss briefly.

Sometimes it may come in our mind that why water is continuously moving on the surface. To find out the answer to this question, we must understand the basic definition of the fluid. **Fluid** is a substance that has no distinct structure and easily responds to pressure externally. Continuous deformation occurs in fluids; therefore, they never stop flowing.

Similarly, Bernoulli's equation can be viewed as an expression of mechanical energy balance. We assumed that in regions of inviscid fluid, compressibility and frictional effects are ignored and the Bernoulli's equation is a mathematical expression where sum of the kinetic energy, potential energy, and the pressure energy of a fluid particle along the

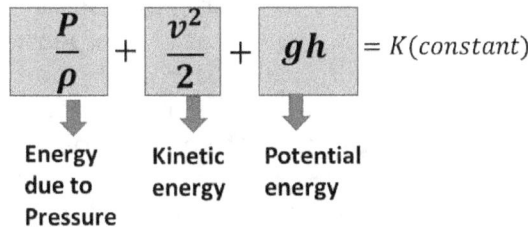

$$\frac{P}{\rho} + \frac{v^2}{2} + gh = K(constant)$$

Energy due to Pressure   Kinetic energy   Potential energy

streamline remains constant for the steady flow through a

streamline. Therefore, **Bernoulli's equation** known as the "conservation of mechanical energy." Recall the fact that energy can neither be created nor destroyed, however energy can only be converted from one form of energy to another. In other words, amount of the energy in working system remains same unless a force is applied to the system from a distance.

So that Bernoulli's equation can be expressed as *"the work done by the pressure and gravity forces on the fluids is equal to the increase in the kinetic energy of the fluid particles."*

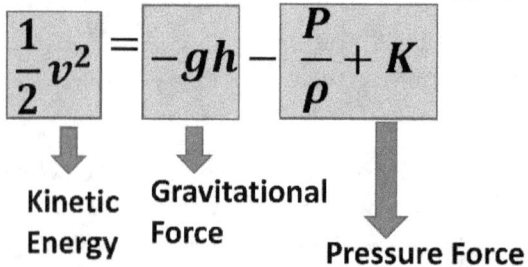

$$\frac{1}{2}v^2 = -gh - \frac{P}{\rho} + K$$

Kinetic Energy   Gravitational Force   Pressure Force

Moreover, **Bernoulli's equation** which will discuss vividly later on. Let us quickly look at its applications.

APPLICATIONS OF BERNOULLI'S EQUATION

1. WORKING OF THE SPRAYER

When we squeeze the balloon shaped piston produces the stream of air with high velocity tends to cross the point C.

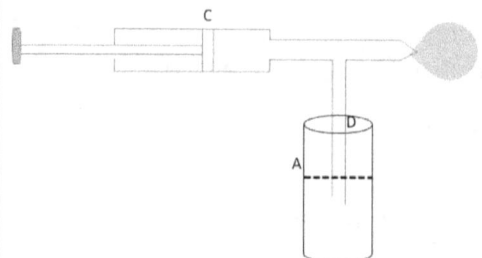

So, according to Bernoulli's principle, air have less pressure near the point C because of high velocity of the air near the point C. However, the other end of the tube is immersed in the liquid such that there is high pressure at the point D in the tube because open free surface of liquid has minimal velocity. The difference in pressure in the region of tube which enable the liquid to carried away by the stream of air for spraying purpose.

1. **Working of Bunsen burner**

   When a gas passes through a nozzle N (shown in Figure), the velocity of the gas increases and the pressure inside the vertical pipe decreases.

   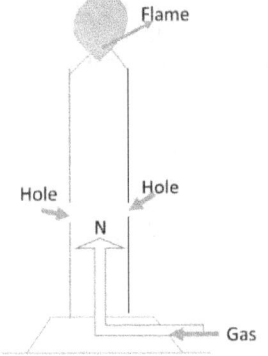

   So, gas flowing through the nozzle has low pressure than the steady air surrounding it. This variation in pressure causes air rushes into the burner through the side holes from outside by venturi effect. As a result, mixture of gas and air moves upward which burns at the top of the burner.

2. BLOWING THE ROOF DURING STORM

During the wind storm, the speed of air just above the roof is much high. So, by Bernoulli's principle the pressure just above the roof is less than the pressure below the roof so results an upward force due to the pressure gradient acts on the roof that's makes the roof blown off without damaging other parts of the house.

Source from www.dreamstime.com

3. LIFT OF THE WINGS OF AIRPLANE

The shape of the airplane's wings are such that it is slightly convex upward and concave downward as shown in figure. Thus, the air passing over the top of air wings travels a longer distance than that the lower surface at instant time. So, the velocity of the air above the wings are much high than the surface below the wings. By Bernoulli's theorem, the pressure on the upper surface is less as compare to the pressure on the lower surface of the wing. This pressure difference provides an additional thrust (T) on the wing. The vertical component of the thrust gives the lift (L) and the horizontal component gives the forward drag (D) to the wings.

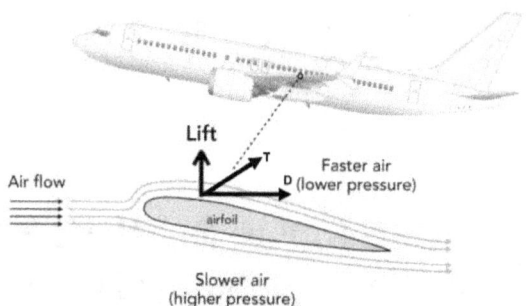

4. DANGEROUS TO STAND ON THE EDGE OF A PLATFORM

When a fast-moving train cross a person standing near railway track, the person has a tendency to fall towards the train. This happen due to the reason that a fast-moving train produces large velocity in air between the person and the train and pressure decreases by Bernoulli's theorem. Thus, the excess pressure pushes the person towards the train. So, it is advised not to stand on the edge on the platform while a fast moving-train crosses the platform.

5. MAGNUS EFFECT

Consider a ball moving to the right which given a spin at the top of the ball as shown in Figure. It has been observed that the velocity of the air on the upper surface of the ball is higher than the lower surface (below the ball). Now, recall the Bernoulli's theorem stated that the increases in velocity decline in pressure on moving plane. So, we can say that pressure on the upper surface (air on the top of the ball) is less than the pressure on lower surface (air below the ball) which results in a net upward force on the spinning ball which possible to move in curved path. This phenomenon is known as "Magnus effect."

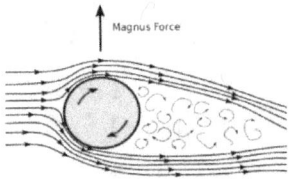

6. SPEED OF EFFLUX IN THE TANK

Consider a fluid of density ρ in an open surface tank up-to the level A. Let there be a small hole (orifice) in the tank at B at a depth h from the surface A as shown in figure.

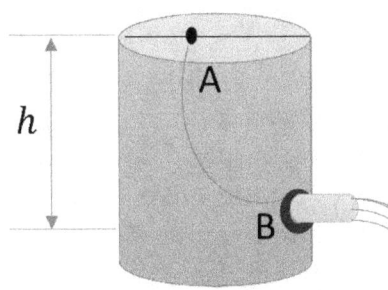

Also, let the fluid motion which comes out from the orifice be v (also called as speed of efflux). Since hole is very small as compared to tank size, so we can take the velocity of fluid at A as zero. Also, both the points A and B are exposed to atmosphere then, we have the pressure at A and B will be equal to atmospheric pressure (P_{atm}).

Now, we apply the Bernoulli's equation along a streamline from point A to point B and we obtain as

$$\frac{P_A}{\rho g} + \frac{v_A^2}{2g} + h_A = \frac{P_B}{\rho g} + \frac{v_B^2}{2g} + h_B$$

$$\Rightarrow \frac{P_{atm}}{\rho g} + \frac{v_A^{2\,0}}{2g} + h = \frac{P_{atm}}{\rho g} + \frac{v_B^2}{2g} + h_B^{\,0}$$

$$\Rightarrow \frac{v^2}{2g} = h \Rightarrow v = \sqrt{2gh}$$

In this way we can calculate the speed of efflux which is equivalent to velocity of an object freely falling from a certain height h. This phenomenon is also known as **"Torricelli's Law."**

2. BERNOULLI'S EQUATION
APPLICATIONS

PRESSURE DROP IN FLOWS THROUGH A PIPE

EXAMPLE 1. Water enters a horizontal pipe of non-uniform cross-section with a speed of $0.7 \, \text{m s}^{-1}$ with the pressure of $1000 \, \text{N m}^{-2}$ and emerge out from the other end with a speed of $0.4 \, \text{m s}^{-1}$. Determine the pressure at the outlet of the pipe provided density of the water flowing through pipe is $1000 \, \text{kg m}^{-3}$.

Solution: This problem involves the conversion of flow, kinetic and potential energies to each other without using any pumps, turbines, and devices. Here, we take point P and point Q at the inlet and outlet of the pipe as shown in figure. The difference in pressure between point P and the point Q (at the outlet of the pipe) possible for flowing the fluid (water) in the pipe. The pressure at the outlet of the pipe can be determined on consider following assumptions:

1. The water flow which exit in the air is steady, incompressible, and irrotational.
2. The friction between the water and air has ignored.
3. The surface tension effects has ignored.
4. The irreversibility effect that occur at the outlet of the pipe because of abrupt contraction will not consider.

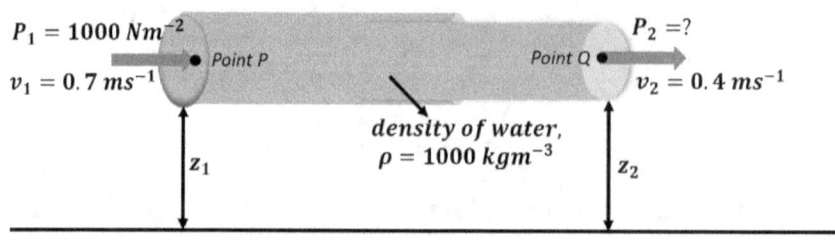

$P_1 = 1000 \, Nm^{-2}$ Point P   Point Q $P_2 = ?$

$v_1 = 0.7 \, ms^{-1}$ $v_2 = 0.4 \, ms^{-1}$

density of water, $\rho = 1000 \, kgm^{-3}$

z_1 z_2

Reference level

Now, applying the Bernoulli's equation along the streamline from the point P to the point Q, we obtain as:

$$\frac{P_1}{\rho g} + \frac{v_1^2}{2g} + z_1 = \frac{P_2}{\rho g} + \frac{v_2^2}{2g} + z_2 \quad \text{-----(1)}$$

Since water past through a horizontal pipe, we have

$$z_1 = z_2$$

Thus, equation (1) reduced to

$$\frac{P_1}{\rho g} + \frac{v_1^2}{2g} = \frac{P_2}{\rho g} + \frac{v_2^2}{2g}$$

$$\Rightarrow \frac{P_1}{\rho} + \frac{v_1^2}{2} = \frac{P_2}{\rho} + \frac{v_2^2}{2}$$

$$\Rightarrow \frac{1}{2}\left(v_1^2 - v_2^2\right) = \frac{1}{\rho}\left(P_2 - P_1\right)$$

$$\Rightarrow P_2 = \frac{\rho}{2}\left(v_1^2 - v_2^2\right) + P_1 = \frac{1000}{2}\left[(0.4)^2 - (0.7)^2\right] + 1000$$

$$\Rightarrow P_2 = 835 \text{ N m}^{-2}$$

Discussion: On ignoring the frictional effect, we found that the pressure at the outlet of the pipe is 835 N m^{-2}. However, if we consider the friction between the water and pipe surface along with other irreversible losses, then pressure would be higher than 835 N m^{-2}. Also, it has observe that the pressure at the point Q is lower than atmospheric pressure . In case, if the elevation difference between point P and point Q is high, then the pressure at the point Q fallen more vividly and some water can vapourate in the form of drops on the outlet of the pipe.

WATER EXIT FROM A HOLE OFA TANK

EXAMPLE 2. A tank filled with water up to the height of 5 m whose upper surface is opened. A rounded outlet tap is attached at the bottom of the tank which is now opened and water flows out. Determine the maximum water velocity at the outlet tap.

Solution: This problem involves the conversion of flow,

kinetic and potential energies to each other without using any pumps, turbines, and devices with large frictional losses. The water exit with maximum velocity under the following assumptions:

1. The water exit slowly such that flow can be approximated as steady.
2. Irreversible losses in the outlet are ignored.
3. The fluid motion is incompressible and irrotational.

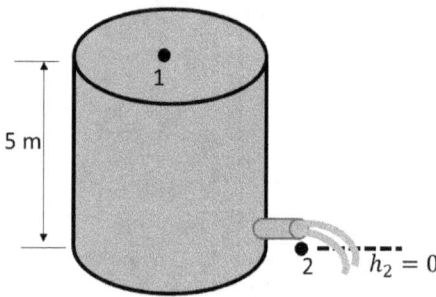

Then, we consider point 1 at the free surface of water on the top of the tank such that $P_1 = P_{atm}$(open to the atmosphere), and point 2 at the center of the outlet tap attached at the bottom of the tank. Then, we consider velocity of water inside the tank taken very small compared to the velocity of exit water from the outlet i.e., $(v_1 \ll v_2)$.Next, we consider the center point of the outlet as the reference level so that we have, $h_2 = 0$ and $h_1 = 5\,m$. Again, $P_2 = P_{atm}$(as the water discharge to the atmosphere from the outlet. Now, we apply the Bernoulli's equation along a streamline from point 1 to point 2 and we obtain as

ignore

$$\cancel{\frac{P_1}{\rho g}} + \cancel{\frac{v_1^2}{2g}} + h_1 = \cancel{\frac{P_2}{\rho g}} + \frac{v_2^2}{2g} + \cancel{h_2}^{\,0}$$

$$\Rightarrow h_1 = \frac{v_2^2}{2g}$$

$$\Rightarrow v_2 = \sqrt{2gh_1} = \sqrt{2 \times 9.81 \tfrac{m}{s^2} \times 5\,m} = 9.9\tfrac{m}{s}$$

Discussion: Thus, the water comes out from the tank with maximum velocity of 9.9 m/s.
If orifice would be sharp-edged instead of round shape, then flow would be disturbed and exit velocity would be less than

9.9 m/s so we can say that the frictional effect and flow disturbance can not be negligible. Also, we knew from the conservation of the mass that

$$\left(\frac{v_1}{v_2}\right)^2 = \left(\frac{d_2}{d_1}\right)^4$$

If we take

$$\frac{d_2}{d_1} = 0.1$$

$$\Rightarrow \left(\frac{v_1}{v_2}\right)^2 = 0.0001 \ll 1$$

So our assumption is justified.

PETROL DISSIPATION FROM FUEL TANK OF A CAR

EXAMPLE 3. On the way to the park, a car runs out of petrol, it would necessary to extract gas out of the fuel tank of the car. The siphon is a small diameter pipe for extraction and insert one end of pipe in the fuel tank so that pipe will fill with petrol via suction and other end of the pipe in the gas can which is below the level of fuel tank.Here, point 2 is positioned 0.75 m below the point 1 and point 3 is 2 m above the point 1. The pipe diameter is 5 mm. Determine the minimum time to extract 4 L of petrol from the fuel tank to the gas can, and the pressure at the point 3 provided density of petrol is 750 kg/m^3.

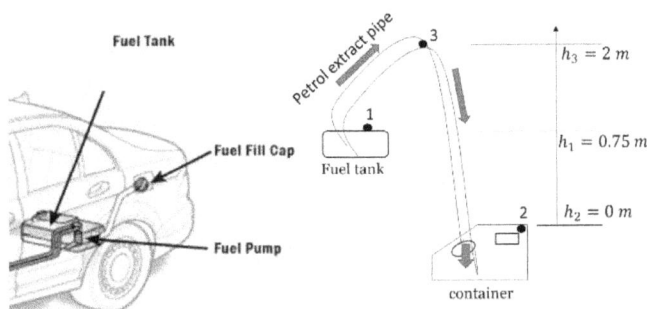

Source from https://in.pinterest.com

Solution: Here, we take point 1 to be free surface of petrol in fuel tank. The difference in pressure between point 1 and the point 2 (at the outlet of the pipe) possible for flowing the fluid in the pipe. The petrol which is to be extract from the fuel tank with the min time can be determined on consider following assumptions:

1. The flow motion is steady and incompressible.
2. The change in the petrol surface level inside the tank is negligible.
3. The frictional loses during the extraction would be ignored.
4. The density of petrol is $750\ kg/m^3$.

Then, we consider point 1 to be free surface of petrol in fuel tank so that $P_1 = P_{atm}$. Also, the size of the fuel tank is large as compared to the pipe diameter such that we take $v_1 = 0$. Point 2 taken as the reference level i.e., $h_2 = 0$ and petrol discharges into the atmosphere so $P_2 = P_{atm}$. Also, we taken into account that $v_2 = v_3$.

Then, we apply the Bernoulli's equation along a streamline from point 1 to point 2 and we obtain as

$$\underset{\text{ignore}}{\cancel{\frac{P_1}{\rho g}} + \cancel{\frac{v_1^2}{2g}}} + h_1 = \cancel{\frac{P_2}{\rho g}} + \cancel{\frac{v_2^2}{2g}} + \cancel{h_2}^{\ 0}$$

$$\Rightarrow h_1 = \frac{v_2^2}{2g}$$

$$\Rightarrow v_2 = \sqrt{2gh_1} = \sqrt{2 \times 9.81\,\frac{m}{s^2} \times 0.75\ m} = 3.84\ \frac{m}{s}$$

The cross-sectional area of the pipe and discharge rate of petrol are given as:

$$A = \frac{\pi d^2}{4} = \frac{\pi}{4}(5 \times 10^{-3}m)^2 = 1.96 \times 10^{-5}\ m^2$$

$$Q = v_2 A = \left(3.84\,\frac{m}{s}\right)(1.96 \times 10^{-5}\ m^2) = 7.53 \times 10^{-5}\frac{m^3}{s}$$

$$\Rightarrow Q = 0.0753\ \frac{L}{s}$$

Then required time for extract 4L of petrol is

$$\Delta t = \frac{V}{Q} = \frac{4\,L}{0.0753\,\frac{L}{s}} = 53.1\ s$$

Next, apply the Bernoulli's equation along a streamline

between point 3 and point 2, we obtain as:

$$\frac{P_2}{\rho g} + \frac{v_2^2}{2g} + \cancel{h_2} = \cancel{\frac{0}{}} \frac{P_3}{\rho g} + \frac{v_3^2}{2g} + h_3$$

$$\Rightarrow \frac{P_{atm}}{\rho g} = \frac{P_3}{\rho g} + h_3$$

$$P_3 = P_{atm} - \rho g h_3$$

$$\Rightarrow P_3 = 101.3 \; kPa - \left(750 \frac{kg}{m^3}\right)\left(9.81 \frac{m}{s^2}\right)(2.75 \; m)\left(\frac{1 \; N}{1 \; kg \frac{m}{s^2}}\right)\left(\frac{1 \; kPa}{10^3 \frac{N}{m^2}}\right)$$

$$\Rightarrow P_3 = 81.1 \; KPa$$

Discussion: On ignoring the frictional effect, we found that the minimum time to extract 4 L of petrol is 53.1 sec. However, if we consider the friction between the petrol and pipe surface along with other irreversible losses, then time would be longer than 53.1 sec. Also, it has observe that the pressure at the point 3 is lower than atmospheric pressure . In case, if the elevation difference between point 1 and point 3 is high, then the pressure at the point 3 fallen more vividly and some petrol can vapourate which can form drops on the top surface of fuel tank.

WATER LEVEL RISES IN A CYCLONE

EXAMPLE 4. The figure depicts a tropical cyclone swelling up over the ocean. The open sea surface level claimed as point 1 being the reference level from where elevation difference upto the eye of the cyclone is 320 km and has atmospheric pressure is 762 mm Hg. Also, the atmospheric pressure at the eye of the cyclone is 560 mm Hg. It has mentioned that density of seawater, mercury and air at sea level are 1025 kg/m^3, 13,600 kg/m^3, and 1.2 kg/m^3 respectively. Determine

1. The elevation at the eye of cyclone from the sea level .

2. The elevation up to the point 2 in region of hurricane from the sea level provided the wind velocity is 250 km/hr.

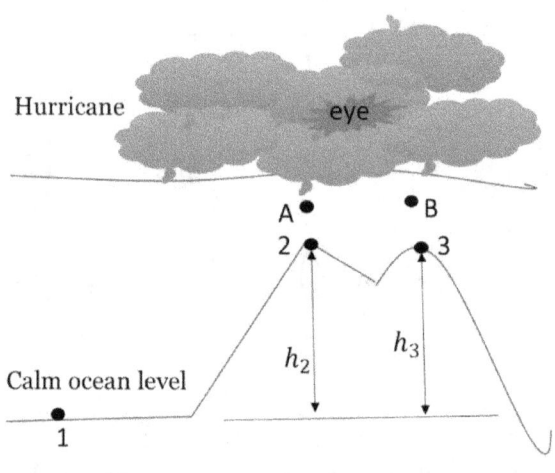

Solution: Reduced atmospheric pressure over the water would be the reason for elevation of storm level. Thus, the point 1 taken as point of reference level so that decreased in pressure between the the point 2 and point 1 would be the reason for elevation of ocean water rise to the point 2. However, certains assumptions should be taken in order to apply the Bernoulli's equation in present case.

1. The airflow within the cyclone is assumed to be steady, incompressible, and irrotational in order to apply Bernoulli's equation.
2. The amount of water sucked into the air during cyclone had been ignored.
3. Given that density of airflow, seawater, and mercury are 1.2 kg/m^3, 1025 kg/m^3, and 1360 kg/m^3.
4. At point 3, motion of airflow would be ignored.

Consider change in pressure would be equal to hydrostatic pressure exerted by mercury which is equivalent to the pressure exerted by seawater on reference level. So we obtain as:

$$\Delta P = (\rho g h)_{Hg} = (\rho g h)_{sea\,water}$$

$$\Rightarrow h_{sw} = \frac{\rho_{Hg}}{\rho_{sw}}\, h_{Hg}$$

16

$$\Rightarrow h_3 = \left(\frac{1360\ Kg/m^3}{1025\ kg/m^3}\right)[(762-560)mm\ Hg]\left(\frac{1\ m}{1000\ mm}\right)$$

$$\Rightarrow h_3 = 2.68\ m$$

Next, applying Bernoulli's equation between point A and point B, which are on the top of point 2 and point 3, respectively. Here, $v_B \approx 0$ and $h_A = h_B$ (both the points are on the same horizontal line). So, we obtain as:

$$\frac{P_A}{\rho g} + \frac{v_A^2}{2g} + h_A = \frac{P_B}{\rho g} + \frac{v_B^2}{2g} + h_B$$

$$\Rightarrow \frac{P_B - P_A}{\rho g} = \frac{v_A^2}{2g}$$

$$\Rightarrow h_{air} = \frac{P_B - P_A}{\rho g} = \frac{v_A^2}{2g} = \left(\frac{\left(250\frac{km}{hr}\right)^2}{2.\left(9.81\frac{m}{s^2}\right)}\right) \times \left(\frac{1\frac{m}{s}}{3.6\frac{km}{hr}}\right)^2$$

$$\Rightarrow h_{air} = 246\ m$$

Next, density of airflow close to the eye of cyclone is given by

$$\rho_{air} = \frac{P_{air}}{P_{atm\ air}}\rho_{atm\ air}$$

$$\Rightarrow \rho_{air} = \left(\frac{560\ mm}{762\ mm}\right)(1.2\ kg/m^3)$$

$$\Rightarrow \rho_{air} = 0.882\ kg/m^3$$

Consider change in pressure would be equal to hydrostatic pressure exerted by air which is equivalent to the pressure exerted by seawater on reference level. So we obtain as:

$$\Delta P = (\rho g h)_{air} = (\rho g h)_{sea\ water}$$

$$\Rightarrow h_4 = \frac{\rho_{air}}{\rho_{sw}}h_{air}$$

$$\Rightarrow h_4 = \left(\frac{0.882\ kg/m^3}{1025\ kg/m^3}\right)(246\ m)$$

$$\Rightarrow h_4 = 0.21 \ m$$

Because of high wind velocity makes the possibility for elevation of sea water of 0.21 m above the sea level. Then, total elevation of height at the point 2 calculated as:

$$h_2 = h_3 + h_4$$

$$\Rightarrow h_2 = 2.68 + 0.21 = 2.89 \ m$$

Discussion: Thus, maximum height of seawater due to high speed wind cannot be more than 0.21 m from the sea level. However, If the cyclonic airflow would be rotational at its eye position then we need to consider the vorticity and other parameters which results on restriction on the applicability of Bernoulli's equation .

MOVEMENT OF FLUID IN A SIPHON

EXAMPLE 5. A manometer and a pitot tube are inserted in a pipe as shown in Figure. Determine the fluid motion at the center of the pipe for the indicated water column heights in devices.

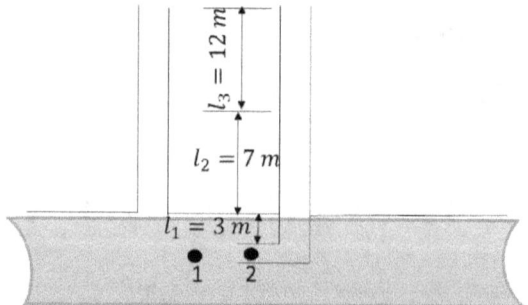

Solution: We take point 1 and point 2 along the streamline at the centerline of the pipe where point 1 is directly under the manometer and point 2 at the tip of pitot tube.

The gauge pressure at the point 1 and point 2 can be expressed as:

$$P_1 = \rho g (l_1 + l_2)$$

$$P_2 = \rho g (l_1 + l_2 + l_3)$$

$$\Rightarrow P_2 - P_1 = \rho g l_3$$

Here, point 1 and point 2 are cose together so that $h_1 = h_2$, and point 2 is a stagnation point and thus $v_2 = 0$. So, applying the Bernoull's equation between the point 1 and point 2 , we obtain as:

$$\frac{P_1}{\rho g} + \frac{v_1^2}{2g} + \cancel{h_1} = \frac{P_2}{\rho g} + \cancel{\frac{v_2^2}{2g}}^{0} + \cancel{h_2}$$

$$\Rightarrow \frac{P_2 - P_1}{\rho g} = \frac{v_1^2}{2g}$$

$$\Rightarrow v_1^2 = \frac{2}{\rho}(P_2 - P_1)$$

$$\Rightarrow v_1^2 = \frac{2}{\rho}(\rho g l_3) = 2g l_3$$

$$\Rightarrow v_1 = \sqrt{2g h_3} = \sqrt{2 \times (9.81 \ ms^2) \times (0.12 \ m)}$$

$$\Rightarrow v_1 = 1.53 \ m/s$$

DISCHARGE OF CHEMICAL THROUGH A SIPHON

EXAMPLE 6. A siphon inserted in a open vessel to discharge HCl chemical having density $0.8 \ kg/m^3$. Given that vapour pressure of HCl is $29.5 \ kPa$ and atmospheric pressure is $101 \ kPa$. Determine following parameters:

1. The fluid motion inside the siphon.
2. The pressure at the point A and point B.
3. The pressure at the point C.
4. The maximum height of HCl in the siphon above the free surface of vessel.
5. The maximum column height of HCl in the right side of

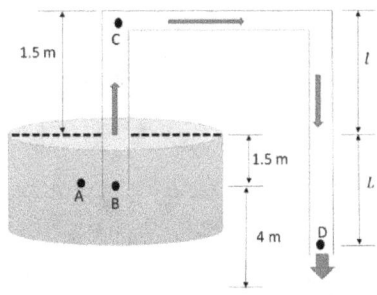

siphon.

Solution: Given figure depicts the discharge of HCl chemical from the open water vessel into atomsphere through a siphon. Here, point A and point B are on the same horizontal plane but point A is outside the siphon and point B is inside the siphon. Then consider point 1 is on the free surface of the vessel filled with water and point C inside the siphon such a way the point D is last point on discharging the chemical into atmosphere. Thus, to calculate the above mentioned parameters, we need to consider following assumptions in oder to apply the Bernoulli's equation

1. The flow motion is steady and incompressible.
2. The change in the chemical surface level inside the vessel is negligible.
3. The frictional loses during the extraction would be ignored.
4. The density of HCl is $0.8 \ kg/m^3$.

Consider the plane through the point D as datum, and siphon discharge the chemical into atmosphere, so at point D the atmospheric pressure holds. Here, area of the vessel is large as compared to the area of siphon so that the inside the vessel is very small compared to the emerging chemical flow $(v_1^2 \ll v_D^2)$ and we consider the elevation just below the free surface as the reference level ($h_1 = 4 + 1.5 = 5.5 \ m$). Now, we apply the Bernoulli's equation along the streamline from point 1 to the point D, we obtain as:

$$\frac{\overset{P_{atm}}{\cancel{P_1}}}{\rho g} + \frac{\overset{ignored}{\cancel{v_1^2}}}{2g} + h_1 = \frac{\overset{P_{atm}}{\cancel{P_D}}}{\rho g} + \frac{v_D^2}{2g} + \overset{0}{\cancel{h_D}}$$

$$\Rightarrow h_1 = \frac{v_D^2}{2g}$$

$$\Rightarrow v_D = \sqrt{2gh_1}$$

$$\Rightarrow v_D = \sqrt{2 \times (9.81\frac{m}{s^2})(5.5 \ m)} = 10.39 \ m/s$$

Next, pressure at the point A is given by
$$P_A = P_{atm} + \rho g h$$

$$\Rightarrow P_A = 101 \ kPa + \left(0.8\frac{kg}{m^3}\right) \times \left(9.81\frac{m}{s^2}\right) \times (1.5 \ m)\left(\frac{1 \ N}{1 \ kg\frac{m}{s^2}}\right)\left(\frac{1 \ kPa}{10^3\frac{N}{m^2}}\right)$$

$$\Rightarrow P_A = 101 \ kPa + 11.77 \ kPa = 112.77 \ kPa$$

Now, we apply the Bernoulli's equation along the streamline from point A to the point B, we obtain as:

$$\frac{P_A}{\rho g} + \frac{v_A^2}{2g}^{\ 0} + h_A = \frac{P_B}{\rho g} + \frac{v_B^2}{2g} + h_B$$

$$\Rightarrow \frac{P_A}{\rho g} + 0 + 4 = \frac{P_B}{\rho g} + \frac{v_B^2}{2g} + 4$$

$$\Rightarrow P_B = P_A - \rho \frac{v_B^2}{2}$$

$$\Rightarrow P_B = 112.77 \ kPa - \left(0.8 \frac{kg}{m^3}\right)\left(\frac{\left(10.39\frac{m}{s}\right)^2}{2}\right)\left(\frac{1 \ N}{1 \ kg \frac{m}{s^2}}\right)\left(\frac{1 \ kPa}{10^3 \frac{N}{m^2}}\right)$$

$$\Rightarrow P_B = 112.77 \ kPa - 43.18 \ kPa = 69.59 \ kPa$$

Now, we apply the Bernoulli's equation along the streamline from point 1 to the point C, we obtain as:

$$\frac{P_1}{\rho g} + \frac{v_1^2}{2g}^{\ 0} + h_1 = \frac{P_C}{\rho g} + \frac{v_C^2}{2g} + h_C$$

$$\Rightarrow \frac{P_{atm}}{\rho g} + 0 + 5.5 = \frac{P_C}{\rho g} + \frac{v_D^2}{2g} + 7$$

$$\Rightarrow P_C = P_{atm} - \rho \frac{v_D^2}{2} + \rho g(5.5 - 7)$$

$$\Rightarrow P_C = 101 \ kPa - \left(0.8 \frac{kg}{m^3}\right)\left(\frac{\left(10.39\frac{m}{s}\right)^2}{2}\right)\left(\frac{1 \ N}{1 \ kg \frac{m}{s^2}}\right)\left(\frac{1 \ kPa}{10^3 \frac{N}{m^2}}\right)$$

$$+ \left(0.8 \frac{kg}{m^3}\right)\left(9.81 \frac{m}{s^2}\right)(-1.5 \ m)\left(\frac{1 \ N}{1 \ kg \frac{m}{s^2}}\right)\left(\frac{1 \ kPa}{10^3 \frac{N}{m^2}}\right)$$

$$\Rightarrow P_C = 101 \ kPa - 43.18 \ kPa - 11.76 \ kPa$$

$$\Rightarrow P_C = 46.06 \ kPa$$

Consider the maximum column height of HCl up to the point C where vapour pressure holds. Next, we apply the Bernoulli's equation along the streamline between point 1 and point C, we obtain as:

$$\frac{P_1}{\rho g} + \frac{v_1^2}{2g}{}^{\,0} + h_1 = \frac{P_C}{\rho g} + \frac{v_C^2}{2g} + h_c$$

$$\Rightarrow \frac{P_{atm}}{\rho g} + 0 + 5.5 = \frac{P_{vp}}{\rho g} + \frac{v_D^2}{2g} + (5.5 + l)$$

$$\Rightarrow l = (P_{atm} - P_{vp})\frac{1}{\rho g} - \frac{v_D^2}{2g}$$

$$\Rightarrow l = \frac{(101 - 29.5)kPa}{\left(0.8\frac{kg}{m^3}\right)\left(9.81\frac{m}{s^2}\right)\left(\frac{1\,N}{1\,kg\frac{m}{s^2}}\right)\left(\frac{1\,kPa}{10^3\frac{N}{m^2}}\right)} - \frac{\left(10.39\frac{m}{s}\right)^2}{2\left(9.81\frac{m}{s^2}\right)}$$

$$\Rightarrow l = 9.11\,m - 5.5\,m = 3.61\,m$$

Again we apply the Bernoulli's equation along the streamline from the point 1 to the point C, we obtain as:

$$\frac{P_1}{\rho g} + \frac{v_1^2}{2g}{}^{\,0} + h_1 = \frac{P_C}{\rho g} + \frac{v_C^2}{2g}{}^{\,ignore} + h_c$$

$$\Rightarrow \frac{P_{atm}}{\rho g} + 0 + L = \frac{P_{vp}}{\rho g} + (1.5 - L)$$

$$\Rightarrow 2L = (P_{atm} - P_{vp})\frac{1}{\rho g} - 1.5$$

$$\Rightarrow 2L = \frac{(101 - 29.5)kPa}{\left(0.8\frac{kg}{m^3}\right)\left(9.81\frac{m}{s^2}\right)\left(\frac{1\,N}{1\,kg\frac{m}{s^2}}\right)\left(\frac{1\,kPa}{10^3\frac{N}{m^2}}\right)} - 1.5\,m$$

$$\Rightarrow 2L = 9.11\,m - 1.5\,m = 7.61\,m$$

FORCE APPLIED ON FLOW THROUGH A PIPE

EXAMPLE 7. Determine the amount of force exerted on the pipe provided the diameter of two ends are 24 m and 12 m. Given that pressure at the larger end of the pipe is 50 kPa and water flows with speed of 8 m/sec.

Solution: This problem involves the conversion of flow, kinetic and potential energies to each other without using any pumps, turbines, and devices with large frictional losses. The water flow which exit in the air is steady, incompressible, and irrotational. . We need to consider following assumptions in

order to apply Bernoulli's equation to the problem.
1. The friction between the water and air has ignored.
2. The surface tension effects has ignored.
3. The irreversibility effect that occur at the outlet of the pipe because of abrupt contraction will not consider.
4. Density of the water is $1000 \ kg/m^3$.

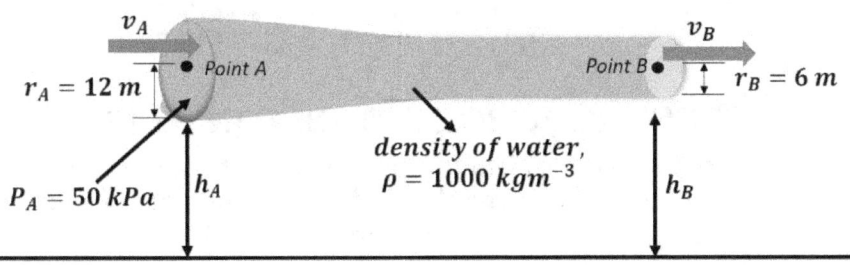

Reference level

Let P_A and P_B be the pressure on the two ends of the pipe. Then, v_A and v_B be the velocities at the two points A and B. then, A_1 and A_2 are the area of cross-sections of two ends of the pipe as shown in figure. Then we have
$$A_1 = \pi r^2 = \pi(12)^2 \ m^2 \ \text{and} \ A_2 = \pi(6)^2 \ m^2$$
From the conservation of mass, we have
$$A_1 v_A = A_2 v_B$$
$$\Rightarrow \pi(12)^2 v_A = \pi(6)^2 v_B$$
$$\Rightarrow v_B = 4v_A$$
Now, applying Bernoulli's equation along the streamline from the point A to the point B, we obtain as:
$$\frac{P_A}{\rho g} + \frac{v_A^2}{2g} + h_A = \frac{P_B}{\rho g} + \frac{v_B^2}{2g} + h_B$$
$$\Rightarrow \frac{P_A}{\rho g} + \frac{v_A^2}{2g} = \frac{P_B}{\rho g} + \frac{v_B^2}{2g}$$
$$[\because h_A = h_B \ (observed \ from \ the \ same \ reference \ level)]$$
$$\Rightarrow P_A - P_B = \frac{\rho}{2}(v_B^2 - v_A^2)$$
$$\Rightarrow P_A - P_B = \frac{\rho}{2}(16v_A^2 - v_A^2)$$
$$\Rightarrow P_B = P_A - \frac{15}{2}\rho v_B^2$$
$$Required \ force = P_A A_1 - P_B A_2$$
$$= \pi(12)^2 P_A - \pi(6)^2 P_B$$
$$= 36\pi(4P_A - P_B)$$

$$= 36\pi\left[4P_A - P_A + \frac{15}{2}\rho v_B^2\right]$$

$$= 36\pi\left[3P_A + \frac{15}{2}\rho v_B^2\right]$$

$$= 36\pi\left[3\times(50\ kPa) + \frac{15}{2}\left(1000\frac{kg}{m^3}\right)\left(8\frac{m}{s}\right)^2\left(\frac{1\ N}{1\ kg\frac{m}{s^2}}\right)\left(\frac{1\ kPa}{10^3\frac{N}{m^2}}\right)\right]$$

$$36\pi[150\ kPa + 60\ kPa] = 36\times\pi\times210\ kPa$$
$$= 36\times(3.14)\times(210\ kPa) = 23738.4\ kPa$$

PRESSURE FORCE INDUCED IN A PIPE

EXAMPLE 8. Determine the pressure at the smaller end of the pipe which discharge the water at 40 Lt/sec. Moreover, diameters of both the larger end and smaller end are 0.6 m and 0.3 m along with the pressure at the larger end holds at 400 kPa. The heights of larger end and smaller end of the pipe from the reference level are at 10 m and 6 m as shown in figure.

Solution: This problem involves the conversion of flow, kinetic and potential energies to each other without using any pumps, turbines, and devices with large frictional losses. The water reaches maximum height under the following assumptions:

1. The water flow which exit in the air is steady, incompressible, and irrotational.
2. The friction between the water and air has ignored.
3. The surface tension effects has ignored.
4. The irreversibility effect that occur at the outlet of the pipe because of abrupt contraction will not consider.
5. Density of the water is 1000 kg/m^3.

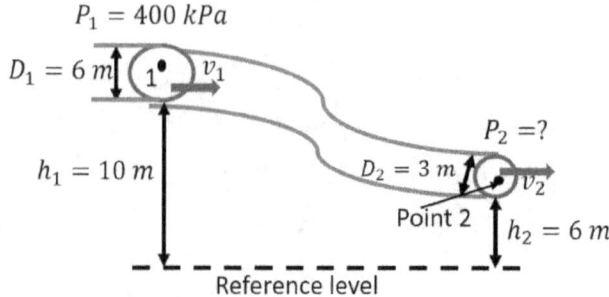

$P_1 = 400\ kPa$

$D_1 = 6\ m$ 1 v_1

$h_1 = 10\ m$

$P_2 = ?$

$D_2 = 3\ m$ v_2

Point 2 $h_2 = 6\ m$

Reference level

Let consider A_1 and A_2 are the area of cross-sections of two ends of the pipe.
$$A_1 = \pi r^2 = \pi(0.3)^2 \ m^2 \quad \text{and} \quad A_2 = \pi(0.15)^2 \ m^2$$

Then consider v_1 and v_2 be the fluid motion at the larger end and smaller end of the pipe.

$$v_1 = \frac{Discharge\ rate}{A_1} = \frac{\left(40\frac{Lt}{s}\right) \times \left(\frac{10^{-3}\ m^3}{1\ Lt}\right)}{\pi\ (0.3)^2\ m^2}$$

$$\Rightarrow v_1 = \frac{\left(0.04\frac{m^3}{s}\right)}{\pi(0.3\ )^2\ m^2} = 0.14\ m/s$$

$$v_2 = \frac{Discharge\ rate}{A_2} = \frac{(0.04\frac{m^3}{s})}{\pi\ (0.15)^2\ m^2} = 0.57\ m/s$$

Then, we consider point 1 at the inlet of the pipe and point 2 at the center of the outlet of the pipe as shown in figure. Then, we consider velocity of water inside the pipe taken very small compared to the velocity of water exit from the outlet i.e., $(v_1 \ll v_2)$. Now, we apply the Bernoulli's equation along a streamline from point 1 to point 2 and we obtain as

$$\frac{P_1}{\rho g} + \frac{v_1^2}{2g} + h_1 = \frac{P_2}{\rho g} + \frac{v_2^2}{2g} + h_2$$

$$\Rightarrow P_2 = P_1 + \frac{\rho}{2}(v_1^2 - v_2^2) + \rho g(h_1 - h_2)$$

$$P_2 = 400\ kPa + \frac{1}{2}\left(10^3\frac{kg}{m^3}\right)\left[(0.14)^2 - (0.57)^2\frac{m^2}{s^2}\right]\left(\frac{1\ N}{1\ kg\frac{m}{s^2}}\right)\left(\frac{1\ kPa}{10^3\frac{N}{m^2}}\right)$$

$$+ \left(10^3\frac{kg}{m^3}\right)\left(9.81\frac{m}{s^2}\right)[(10-6)\ m]\left(\frac{1\ N}{1\ kg\frac{m}{s^2}}\right)\left(\frac{1\ kPa}{10^3\frac{N}{m^2}}\right)$$

$$\Rightarrow P_2 = 400\ kPa + 0.152\ kPa + 39.24\ kPa$$

$$\Rightarrow P_2 = 439.392\ kPa$$

Discussion: On ignoring the frictional effect, we found that pressure at the smaller end of the pipe is 439.392 kPa. However, if we consider the friction between the water and pipe surface along with other irreversible losses, then rate of discharge will decreases. In case, if the elevation difference between point 1 and point 2 is high, then the pressure at the point 2 fallen more vividly and some water can vapourate in the form of drops on the outlet of the pipe.

LIFT FORCE INDUCED IN AEROPLANE WINGS

EXAMPLE 9. Air is flowing past a horizontal air plane wing such that its fluid motion is 120 m/s over the upper surface and 90 m/s over the lower surface. If the density of air is 1.3 kg/m^3 and wing is 10 m long and on average width of 2 m. Determine the total lift of the wing of the aeroplane.

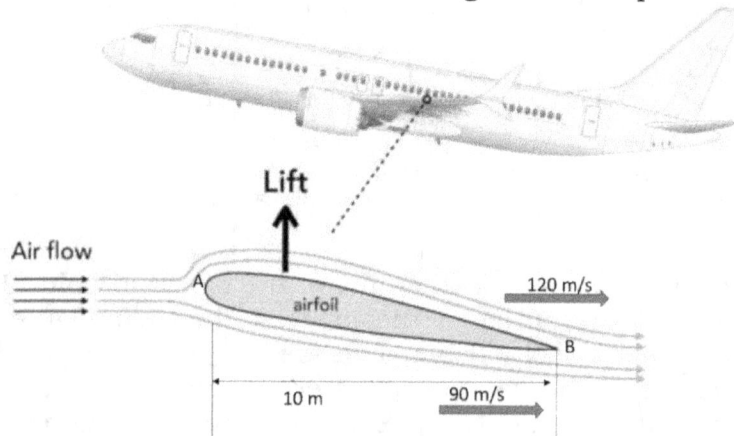

Solution: This problem involves the conversion of flow, kinetic and potential energies to each other without using any pumps, turbines, and devices with large frictional losses. We need to consider following assumptions in order to apply Bernoulli's equation to the problem.

1. The flow motion is steady and incompressible and irrotational.
2. The change in the air level over the wing surface is negligible.
3. The frictional loses during the air flow would be ignored.

Now, applying the Bernoulli's equation along the streamline from the point A to the point B, we obtain as:

$$\frac{P_A}{\rho g} + \frac{v_A^2}{2g} + h_A = \frac{P_B}{\rho g} + \frac{v_B^2}{2g} + h_B$$

Since air is past through the horizontal airplane wing, so

$$h_A = h_B$$

$$\Rightarrow \frac{P_A}{\rho g} + \frac{v_A^2}{2g} = \frac{P_B}{\rho g} + \frac{v_B^2}{2g}$$

$$\Rightarrow \frac{P_A - P_B}{\rho} = \frac{1}{2}(v_B^2 - v_A^2)$$

$$\Rightarrow P_A - P_B = \frac{\rho}{2}\left(v_B^2 - v_A^2\right)$$

$$\Rightarrow P_A - P_B = \frac{1}{2}\left(1.3\,\frac{kg}{m^3}\right)\left(14400 - 8100\,\frac{m^2}{s^2}\right)\left(\frac{1\,N}{1\,kg\,\frac{m}{s^2}}\right)$$

$$\Rightarrow P_A - P_B = 4.095 \times 10^3\,\frac{N}{m^2}$$

$$\Rightarrow Total\ lift\ on\ the\ wing = (P_A - P_B) \times Area\ of\ the\ wing$$

$$\Rightarrow Total\ lift = (4.095 \times 10^3\,N/m^2) \times (10 \times 2\ m^2)$$

$$\Rightarrow Total\ lift = 8.190 \times 10^4\ N$$

Discussion: On ignoring the frictional effect, we found that amount of lift force on airplane wing is 8.190×10^4 N. However, if we consider the friction between the air and surface of airplane wing along with other irreversible losses, then total lift of the wing of the airplane will decrease.

RISE IN WATER LEVELS IN A ROTATING VESSEL

EXAMPLE 10. A fluid is kept in a opened cylindrical vessel which is rotated along its axis. Determine the difference in heights of the fluid from the center of the vessel to its sides. If the radius of the vessel is 0.05 m and the speed of revolution is 2 rev/s.

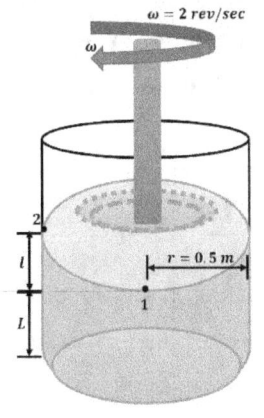

Solution: This problem involves the conversion of flow, kinetic and potential energies to each other without using any pumps, turbines, and devices with large frictional losses. We need to consider following assumptions in order to apply Bernoulli's equation to the problem.

4. The flow motion is steady and incompressible.

5. The change in the surface level inside the vessel is negligible.

6. The frictional loses during the rotation would be ignored.

Given that the fluid motion is rotational so we have

$$v_1 = r \times \Omega = r \times 2\pi\omega$$

$$\Rightarrow v_1 = 0.05\ m \times 2 \times 3.14 \times 2\ \frac{1}{s}$$

27

$$\Rightarrow v_1 = 0.628 \, m/s$$

Then we consider point 1 is at the centre of the axis of rotation and point 2 is at elevation level on the wall of cylindrical vessel as shown in figure. Since both the point are on the free surface of fluid so that $P_1 = P_2 = P_{atm}$. Due to no slip condition, velocity of the fluid at the point 2 is almost zero (i.e., $v_2 = 0$). The elevation of fluid level due to the rotation is l m from the center of the vessel and L m is the height of the stationary fluid contain in the vessel. Next, applying the Bernoulli's equation at two points 1 and 2,we get

$$\frac{P_1}{\rho g} + \frac{v_1^2}{2g} + h_1 = \frac{P_2}{\rho g} + \frac{v_2^2}{2g} + h_2$$

$$\Rightarrow \frac{P_{atm}}{\rho g} + \frac{v_1^2}{2g} + L = \frac{P_{atm}}{\rho g} + 0 + (L + l)$$

$$\Rightarrow \frac{v_1^2}{2g} = l$$

$$\Rightarrow l = \frac{\left(0.628 \frac{m}{s}\right)^2}{2 \times \left(9.8 \frac{m}{s^2}\right)} = 0.02 \, m$$

Discussion: On ignoring the frictional effect, we found that difference in heights of the fluid is 0.02 m. However, if we consider the friction between the water and wall surface of cylindrical vessel along with other irreversible losses, then difference in heights of the fluid will increase.

DISCHARGE OF FLUID THROUGH A PIPE

EXAMPLE 11. Determine the amount of discharge of fluid whose density $1.25 \times 10^3 \, kg/m^3$ through a horizontal pipe whose the radii of its end are 0.1 m and 0.04 m. The pressure difference throughout the pipe is 10 kPa.

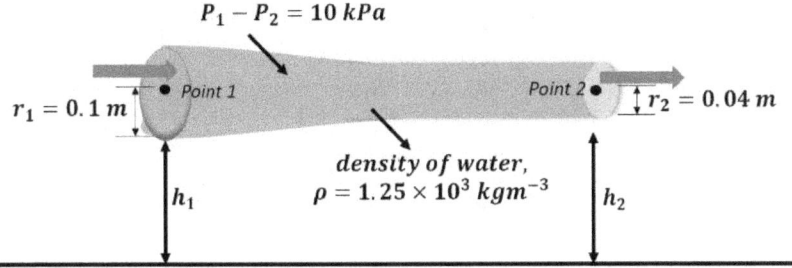

$$P_1 - P_2 = 10 \, kPa$$

Point 1

Point 2

$r_1 = 0.1 \, m$

$r_2 = 0.04 \, m$

h_1

density of water,
$\rho = 1.25 \times 10^3 \, kgm^{-3}$

h_2

Reference level

Solution: This problem involves the conversion of flow, kinetic and potential energies to each other without using any pumps, turbines, and devices with large frictional losses. The amount of discharge of fluid from the pipe can be determine on consider the following assumptions:

1. The fluid motion is steady, incompressible, and irrotational.
2. The friction between the fluid and air has ignored.
3. The surface tension effects has ignored.
4. The irreversibility effect that occur at the outlet of the pipe because of abrupt contraction will not consider.

Then, we consider point 1 at the inlet of the pipe and point 2 at the center of the water streamline emerge out from the outlet of the pipe as shown in figure. Then, we consider velocity of water inside the pipe taken very small compared to the velocity of water exit from the outlet i.e., $(v_1 \ll v_2)$. Now, we apply the Bernoulli's equation along a streamline from point 1 to point 2 and we obtain as

$$\frac{P_1}{\rho g} + \frac{v_1^2}{2g} + h_1 = \frac{P_2}{\rho g} + \frac{v_2^2}{2g} + h_2$$

Since fluid is past through the horizontal pipe, so

$$h_1 = h_2$$

$$\Rightarrow \frac{P_1}{\rho g} + \frac{v_1^2}{2g} = \frac{P_2}{\rho g} + \frac{v_2^2}{2g}$$

$$\Rightarrow \frac{P_1 - P_2}{\rho} = \frac{1}{2}(v_2^2 - v_1^2)$$

$$\Rightarrow v_2^2 - v_1^2 = \frac{2}{\rho}(P_1 - P_2) - - - - - (1)$$

Now, according to the conservation of mass, we have

$$A_1 v_1 = A_2 v_2$$
$$\Rightarrow \frac{v_1}{v_2} = \frac{A_2}{A_1}$$
$$\Rightarrow \frac{v_1}{v_2} = \frac{\pi\, r_2^2}{\pi\, r_1^2} = \frac{r_2^2}{r_1^2} = \frac{(0.04)^2}{(0.1)^2} = 16 \times 10^{-2}$$
$$\Rightarrow v_1 = (16 \times 10^{-2})\, v_2$$

Then, equation (1) becomes

$$v_2^2 - 256 \times 10^{-4}\, v_2^2 = 16 \times 10^{-3}$$
$$\Rightarrow v_2^2 \left[1 - \frac{256}{10^4} \right] = 16 \times 10^{-3}$$
$$\Rightarrow v_2^2 \left(\frac{9744}{10^4} \right) = 16 \times 10^{-3}$$
$$\Rightarrow v_2^2 = \frac{160}{9744} \Rightarrow v_2 = 0.13 \; m/s$$

The amount of discharge of honey is given by

$$= A_2 v_2 = \pi\, r_2^2\, v_2$$
$$\Rightarrow Discharge = 3.14 \times (0.04)^2 \times 0.13$$
$$\Rightarrow Discharge = 6.53 \times 10^{-4} \; m^3/s$$

Discussion: On ignoring the frictional effect, we found that the rate of discharge of water from the horizontal pipe is $6.53 \times 10^{-4} \; m^3/sec$. However, if we consider the friction between the water and pipe surface along with other irreversible losses, then rate of discharge will decreases. In case, if the elevation difference between point 1 and point 2 is high, then the pressure at the point 2 fallen more vividly and some water can vapourate in the form of drops on the outlet of the pipe.

WATER FLOWS OUT THROUGH A SMALL ORIFICE

EXAMPLE 12. Water is flowing from a tap having diameter 0.008 m and water emerge out at a speed of 0.4 m/s. Calculate the diameter of the water stream at a distance 0.2 m below the tap.

Solution: This problem involves the conversion of flow, kinetic and potential energies to each other without using any pumps, turbines, and devices with large frictional losses. The required diameter of water stream can be determine on consider the following assumptions:

1. The water flow which exit in the air is steady, incompressible, and irrotational.
2. The friction between the water and air has ignored.
3. The surface tension effects has ignored.
4. The irreversibility effect that occur at the outlet of the tap because of abrupt contraction will not consider.

Then, we consider point 1 at the free surface of water on the tap of the container such that $P_1 = P_{atm}$ (open to the atmosphere), and point 2 at the center of the water streamline emerge out from the tap as shown in figure. Then, we consider velocity of water inside the tank taken very small

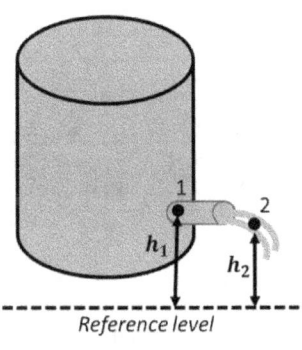

Reference level

compared to the velocity of exit water from the outlet i.e., $(v_1 \ll v_2)$ .Next, $P_2 = P_{atm}$ (as the water discharge to the atmosphere from the tap. Now, we apply the Bernoulli's equation along a streamline from point 1 to point 2 and we obtain as

$$\frac{P_1}{\rho g} + \frac{v_1^2}{2g} + h_1 = \frac{P_2}{\rho g} + \frac{v_2^2}{2g} + h_2$$

$$\Rightarrow \frac{v_2^2}{2} = \frac{v_1^2}{2} + g(h_1 - h_2)$$
$$\Rightarrow v_2^2 = v_1^2 + 2g(h_1 - h_2)$$
$$\Rightarrow v_2^2 = (0.16\frac{m^2}{s^2}) + 2 \times \left(10\frac{m}{s^2}\right) \times (0.2\ m)$$
$$\Rightarrow v_2 = 2.04\ m/s$$

According to law of conservation of mass, we have
$$A_1 v_1 = A_2 v_2$$
$$\Rightarrow \pi\, r_2^2\, v_2 = \pi\, r_1^2\, v_1$$
$$\Rightarrow r_2^2 = \frac{r_1^2\, v_1}{v_2}$$
$$\Rightarrow r_2^2 = \frac{0.16 \times 0.4}{2.04} = 3.14 \times 10^{-6}\ m^2$$
$$\Rightarrow r_2 = 1.77 \times 10^{-3}\ m$$
$$\Rightarrow Diameter = 2\, r_2 = 2 \times 1.77 \times 10^{-3} = 3.54 \times 10^{-3}\ m$$

Discussion: On ignoring the frictional effect, we found that the boundary of water stream is of 3.54×10^{-3} m . However, if we consider the friction between the water and tap surface along with other irreversible losses, then radius of boundaries of water stream would be small as compared to calculated value of radius of the boundary. In case, if the elevation difference between point 1 and point 2 is high, then the pressure at the point 2 fallen more vividly and some water can vapourate in the form of drops on the outlet of the tap.

WATER FLOW FROM A HOSE PIPE

EXAMPLE 13. Water is coming out from a hose pipe and a child place his thumb to cover most of the pipe outlet so that a high-speed flow of water comes out in the form of a thin jet. Determine the maximum height that emerging jet flow could achieve when pipe would hold in upward direction and it is given that the pressure in the pipe is $400\ kPa$

Solution: This problem involves the conversion of flow, kinetic and potential energies to each other without using any pumps, turbines, and devices with large frictional losses. The water reaches maximum height under the following assumptions:

1. The water flow which exit in the air is steady, incompressible, and irrotational.
2. The friction between the water and air has ignored.
3. The surface tension effects has ignored.
4. The irreversibility effect that occur at the outlet of the pipe because of abrupt contraction will not consider.
5. Density of the water is $1000\ kg/m^3$.

The velocity inside the pipe is very small compared to the emerging jet flow $(v_1^2 \ll v_j^2)$ and we consider the elevation just below the pipe outlet as the reference level $(h_1 = 0)$. At the peak of the trajectory flow jet, atmospheric pressure holds and $v_2 = 0$. Now, we apply the Bernoulli's equation along a streamline from point 1 to point 2 and we obtain as

$$\frac{P_1}{\rho g} + \overset{\text{ignore}}{\cancel{\frac{v_1^2}{2g}}} + \cancel{h_1}^{\,0} = \frac{P_2}{\rho g} + \cancel{\frac{v_2^2}{2g}}^{\,0} + h_2$$

$$\Rightarrow \frac{P_1}{\rho g} = \frac{P_{atm}}{\rho g} + h_2$$

$$\Rightarrow h_2 = \frac{P_1 - P_{atm}}{\rho g}$$

$$\Rightarrow h_2 = \frac{400\ kPa}{\left(1000\ \frac{kg}{m^3}\right)\left(9.81\ \frac{m}{s^2}\right)} \times \left(\frac{1000\ \frac{N}{m^2}}{1\ kPa}\right) \times \left(\frac{1\ kg \cdot \frac{m}{s^2}}{1\ N}\right)$$

$$\Rightarrow h_2 = 40.8\ m$$

Discussion: Thus, the emerging water flow jet can reach the maximum height up to 40.8 m into the sky in this case. So, water cannot be elevated further more beyond the height of 40.8 m and all elevations happened under this height only.

PRESSURE REQUIRED TO PUMP THE BLOOD

Example 14. Calculate the minimum pressure required to pump the blood from the heart to the top of the head where vertical elevation is 0.5 m. It is given that the density of human blood is 1.025 kgm^{-3}.

Answer: This problem involves the conversion of flow, kinetic and potential energies to each other without using any pumps, turbines, and devices. Here, we take point P and point Q at the heart and top of the head as shown in figure. The difference in pressure between point P and the point Q possible for flowing the fluid (blood) in the artery.

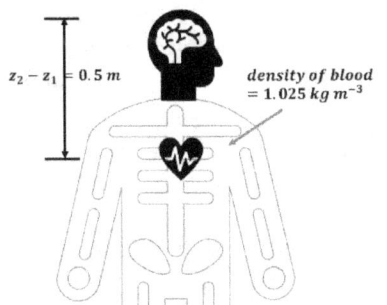

$z_2 - z_1 = 0.5\ m$

density of blood
= 1.025 kg m^{-3}

The pressure required to pump the blood can be determined on consider following assumptions:

1. The blood flow is steady, incompressible, and irrotational.

2. The friction between the blood and artery surface has ignored.

3. The surface tension effects has ignored.

4. The irreversibility effect that occur at the outlet of the artery because of abrupt contraction will not consider.

Now, apply the Bernoulli's equation along the streamline from the point P to the point Q, we obtain as:

$$\frac{P_1}{\rho g} + \frac{v_1^2}{2g} + z_1 = \frac{P_2}{\rho g} + \frac{v_2^2}{2g} + z_2$$

$$\Rightarrow P_1 + \frac{1}{2}\rho v_1^2 + \rho g z_1 = P_2 + \frac{1}{2}\rho v_2^2 + \rho g z_2$$

$$\Rightarrow P_1 - P_2 = \rho g\left(z_2 - z_1\right) + \frac{1}{2}\rho\left(v_2^2 - v_1^2\right) \quad \text{------(1)}$$

In this case, $v_2 = v_1$. Then equation (1) reduced to

$$P_1 - P_2 = \rho g\left(z_2 - z_1\right)$$

Here, $g = 9.8 \text{ ms}^{-2}$

$\rho = 1.025 \text{ kgm}^{-3}$ and $\left(z_2 - z_1\right) = 0.5$ m . Then we have,

$$P_1 - P_2 = 1.025 \times 9.8 \times 0.5 = 5.0225 \text{ Pa}$$

Discussion: On ignoring the frictional effect, we found that the minimum force required to pump the blood from the heart to the top of head is 5.0225 Pa. However, if we consider the friction occur during the blood flow along with other irreversible losses, then amount of force to pump out the blood from the heart will be increases. Also, if the height of the vertical elevation reduces, then less force will be required to pump the blood from the heart.

WATER LEVEL IN TANK WHICH HAS A HOLE

Example 15. A cylindrical tank filled with water up-to the level of 1.6 m. as shown in figure. A hole is made on one of the wall at a depth of 1 m below the water level. Determine at what distance from the bottom of the tank does the emerging stream of water strikes the floor.

Answer: This problem involves the conversion of flow, kinetic and potential energies to each other without using any pumps, turbines, and devices with large frictional losses. The water exit with maximum velocity under the following assumptions:

1. The water exit slowly such that flow can be approximated

as steady.

2. Irreversible losses in the outlet are ignored.

3. The fluid motion is incompressible and irrotational.

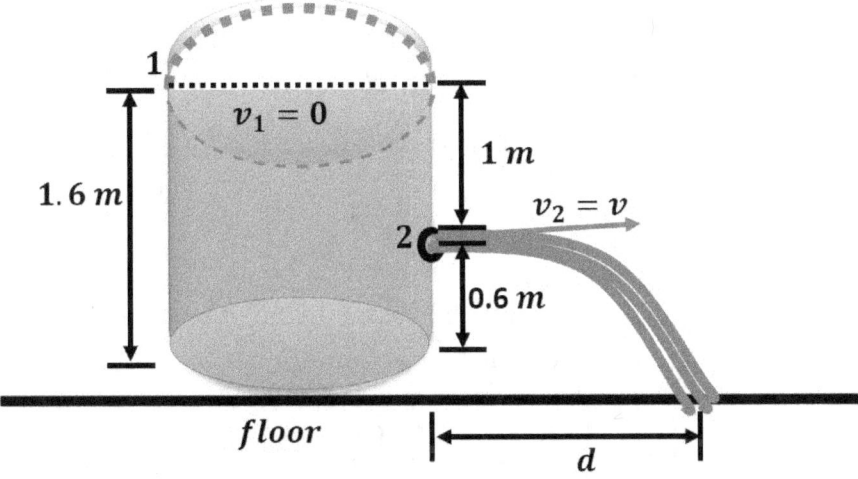

Then, we consider point 1 at the free surface of water on the top of the tank such that $P_1 = P_{atm}$(open to the atmosphere), and point 2 at the center of the outlet hole attached at the bottom of the tank. Then, we consider velocity of water inside the tank taken very small compared to the velocity of exit water from the outlet i.e., $(v_1 \ll v_2)$. In this case, $v_1 = 0$. Again, $P_2 = P_{atm}$(as the water discharge to the atmosphere from the outlet. Now, we apply the Bernoulli's equation along a streamline from point 1 to point 2 and we obtain as

$$\frac{P_1}{\rho g} + \frac{v_1^2}{2g} + z_1 = \frac{P_2}{\rho g} + \frac{v_2^2}{2g} + z_2$$

$$\Rightarrow P_1 + \frac{1}{2}\rho v_1^2 + \rho g z_1 = P_2 + \frac{1}{2}\rho v_2^2 + \rho g z_2 \quad \text{-----(1)}$$

In this case, $P_1 = P_2 = P_{atm}$

$v_1 = 0, v_2 = v$

$z_1 = 1.6$ m, $z_2 = (1.6-1)$ m $= 0.6$ m Then equation 1 reduced to

$$P_{atm} + 0 + \rho g(1.6) = P_{atm} + \frac{1}{2}\rho v^2 + \rho g(0.6)$$

$$\Rightarrow \rho g(1.6-0.6) = \frac{1}{2}\rho v^2$$

$$\Rightarrow \rho g(1) = \frac{1}{2}\rho v^2$$

$$\Rightarrow v = \sqrt{2g(1)} = \sqrt{2 \times 9.8} = 4.43 \text{ m s}^{-1}$$

Let t be the time taken by the water to fall through depth of 0.6 m. From Newton's equation of motion, we have

$$0.6 = 0 + \frac{1}{2}gt^2 \Rightarrow t = \sqrt{\frac{2 \times 0.6}{9.8}} = \sqrt{0.122} = 0.35 \text{ sec}$$

Let d be the distance from the bottom of the tank does the emerging stream of water strikes the floor and it is defined as

$$d = vt = 4.43 \times 0.35 \text{ m} = 1.5505 \text{ m}$$

Discussion: Thus, the water comes out from the tank with maximum velocity of 4.43 m s^{-1}. If hole would be sharp-edged instead of round shape, then flow would be disturbed and exit velocity would be less than 4.43 m s^{-1} so we can say that the frictional effect and flow disturbance cannot be negligible.

PRESSURE FORCE REQUIRED FOR WATER SUPPLY

Example 16. The water flows with the velocity of 3 m/s through the pipe entering the basement whose cross-sectional area is 4×10^{-4} m^2 and pressure at this point is 4×10^5 N/m^2. This pipe tapers to a cross-sectional area of 3×10^{-4} m^2 when it reaches the third floor 31 m above as shown in figure. Determine the pressure on the pipe supply water to the third floor.

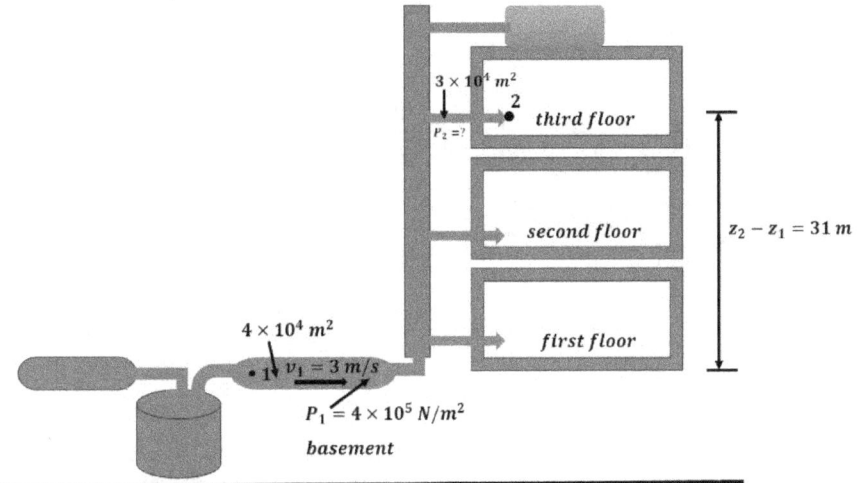

Answer: This problem involves the conversion of flow, kinetic and potential energies to each other without using any pumps, turbines, and devices with large frictional losses. The water flows with maximum velocity under the following assumptions:

1. The water flow approximated as steady.

2. Irreversible losses in the flow through pipe are ignored.

3. The fluid motion is incompressible and irrotational.

From the continuity equation, we have

$$A_1 v_1 = A_2 v_2$$

$$\Rightarrow v_2 = \frac{A_1 v_1}{A_2}$$

$$\Rightarrow v_2 = \frac{4 \times 10^{-4} \times 3}{3 \times 10^{-4}} = 4 \text{ m/s}$$

Then, we consider point 1 at the inlet of the pipe and point 2 at the center of the water streamline emerge out from the outlet of the pipe as shown in figure. Then, we consider velocity of water inside the pipe taken very small compared to the velocity of water exit from the outlet i.e., $(v_1 \ll v_2)$. Now, we apply the Bernoulli's equation along the streamline from the point 1 to point 2 and we have

$$\frac{P_1}{\rho g} + \frac{v_1^2}{2g} + z_1 = \frac{P_2}{\rho g} + \frac{v_2^2}{2g} + z_2$$

$$\Rightarrow P_1 + \frac{1}{2}\rho v_1^2 + \rho g z_1 = P_2 + \frac{1}{2}\rho v_2^2 + \rho g z_2$$

$$\Rightarrow P_2 = P_1 - \frac{1}{2}\left(v_2^2 - v_1^2\right) - \rho g\left(z_2 - z_1\right) \quad \text{-----(1)}$$

In this case, $P_1 = 4\times10^5$ N/m^2 | $v_1 = 3$ m/s, $v_2 = 4$ m/s

$\rho = 10^3$ kg/m^3 | $g = 9.8$ m/s^2 | $z_2 - z_1 = 31$ m

Then equation 1 reduced to

$$P_2 = 3\times10^4 - \frac{1}{2}\left[(4)^2 - (3)^2\right] - 10^3 \times 9.8 \times 31$$

$$P_2 = 4\times10^5 - 3.5 - 10^3 \times 303.8 = 9.6\times10^4 \text{ N/m}^2$$

Discussion: On ignoring the frictional effect, we found that the pressure at the outlet of the pipe is 9.6×10^4 N/m^2. However, if we consider the friction between the water and pipe surface along with other irreversible losses, then pressure would be higher than 9.6×10^4 N/m^2. Also, it has observe that the pressure at the point 2 is lower than atmospheric pressure . In case, if the elevation difference between point 1 and point 2 is high, then the pressure at the point 2 fallen more vividly and some water can vapourate in the form of drops on the outlet of the pipe.

AIR FLOWS THROUGH HELICOPTER'S WINGS

Example 17. A helicopter of mass 3×10^4 kg has total wing area 400 m^2 and flying horizontally with average speed of 300 m/s. Determine the velocity difference and pressure difference between the lower and upper surfaces of the wings provided the density of air is 1.3 kg/m^3 and $g = 9.8$ m/s^2.

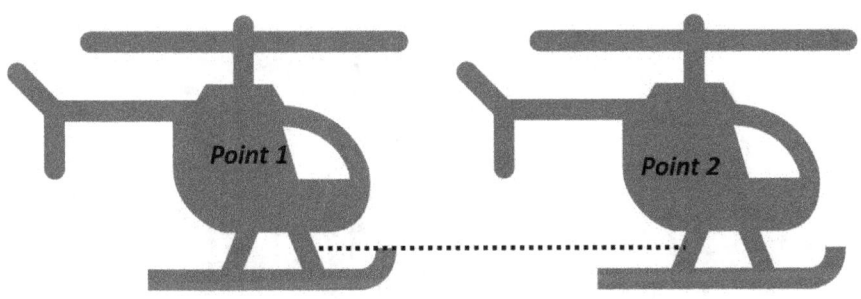

Answer: In this case, mass of helicopter, $M = 3 \times 10^4$ kg

Wing area, $A = 400$ m^2

Acceleration due to gravity, $g = 9.8$ m/s^2

Pressure difference define as

$$P_1 - P_2 = \frac{F}{A} = \frac{Mg}{A} = \frac{3 \times 10^4 \times 9.8}{400}$$

$$\Rightarrow P_1 - P_2 = 735 \text{ N/m}^2$$

Now apply Bernoulli's equation along the streamline from the point 1 to the point 2, we obtain as

$$\frac{P_1}{\rho g} + \frac{v_1^2}{2g} + z_1 = \frac{P_2}{\rho g} + \frac{v_2^2}{2g} + z_2 \quad \text{-----(1)}$$

Since helicopter flying horizontally with average speed of

300 m/s, thus we have

$$z_1 = z_2 \mid \quad \text{average speed} = \frac{v_1 + v_2}{2} = 300 \text{ m/s} \quad \text{Then} \quad \text{equation} \quad (1)$$

reduced to

$$\frac{P_1}{\rho g} + \frac{v_1^2}{2g} = \frac{P_2}{\rho g} + \frac{v_2^2}{2g}$$

$$\Rightarrow \frac{P_1}{\rho} + \frac{v_1^2}{2} = \frac{P_2}{\rho} + \frac{v_2^2}{2}$$

$$\Rightarrow \frac{1}{2}\left(v_2^2 - v_1^2\right) = \frac{1}{\rho}\left(P_1 - P_2\right)$$

$$\Rightarrow v_2 - v_1 = \frac{1}{\rho} \frac{P_1 - P_2}{\left(v_2 + v_1\right)/2}$$

$$\Rightarrow v_2 - v_1 = \frac{1}{1.3}\left(\frac{735}{300}\right) = 1.88 \text{ m/sec}$$

THEORY
DERIVATION OF THE EQUATION

The **Bernoulli's equation** is the mathematical relation between pressure, velocity, and elevation which derivable based on following assumptions. Due to its simplicity, it is widely used in many real-life applications which we will discuss briefly later on in this chapter. So far, now we derive the Bernoulli's equation by integration of Euler's equation of motion, and we illustrate many usefulness and few limitations.

Assumptions: The fluid motion is steady where net frictional forces are negligible. Also, viscous effects are negligible. Since there is no such fluids which has no viscosity, so all fluids have some viscosity. However, in order to derive the Bernoulli's equation, we assume certain regions as inviscid regions of flow (where net viscous or frictional forces are negligibly small compared to the other forces acting on the fluid particles). The fluid motion is irrotational such that the velocity potential exist. The external forces are derivable from the conservative potential function.

Thus, Bernoulli's approximations not adaptable for the flows which are of boundary layers and wakes.

Let ϕ be the velocity potential and F be the force potential. Then by definition we have,

$$u = -\frac{\partial \phi}{\partial x}, \quad v = -\frac{\partial \phi}{\partial y}, \quad w = -\frac{\partial \phi}{\partial z} \quad ----(1)$$

$$X = -\frac{\partial F}{\partial x}, \quad Y = -\frac{\partial F}{\partial y}, \quad Z = -\frac{\partial F}{\partial z} \quad ----(2)$$

Since fluid motion is irrotational, we have

$$\frac{\partial u}{\partial y} = \frac{\partial v}{\partial x}, \quad \frac{\partial v}{\partial z} = \frac{\partial w}{\partial y}, \quad \frac{\partial w}{\partial x} = \frac{\partial u}{\partial z} \quad ---- (3)$$

Next consider the Euler's equations of motion as:

$$\begin{cases} \dfrac{\partial u}{\partial t} + u\dfrac{\partial u}{\partial x} + v\dfrac{\partial u}{\partial y} + w\dfrac{\partial u}{\partial z} = X - \dfrac{1}{\rho}\dfrac{\partial P}{\partial x} \\[2mm] \dfrac{\partial v}{\partial t} + u\dfrac{\partial v}{\partial x} + v\dfrac{\partial v}{\partial y} + w\dfrac{\partial v}{\partial z} = Y - \dfrac{1}{\rho}\dfrac{\partial P}{\partial y} \\[2mm] \dfrac{\partial w}{\partial t} + u\dfrac{\partial w}{\partial x} + v\dfrac{\partial w}{\partial y} + w\dfrac{\partial w}{\partial z} = Z - \dfrac{1}{\rho}\dfrac{\partial P}{\partial z} \end{cases} ---- (4)$$

Manipulating the system of equations given by equ (4) on using the relations given by equations (1), (2), and (3), we obtain,

$$\begin{cases} -\dfrac{\partial^2 \phi}{\partial t \partial x} + u\dfrac{\partial u}{\partial x} + v\dfrac{\partial v}{\partial x} + w\dfrac{\partial w}{\partial x} = -\dfrac{\partial F}{\partial x} - \dfrac{1}{\rho}\dfrac{\partial P}{\partial x} \\[2mm] -\dfrac{\partial^2 \phi}{\partial t \partial y} + u\dfrac{\partial u}{\partial y} + v\dfrac{\partial v}{\partial y} + w\dfrac{\partial w}{\partial y} = -\dfrac{\partial F}{\partial y} - \dfrac{1}{\rho}\dfrac{\partial P}{\partial y} \\[2mm] -\dfrac{\partial^2 \phi}{\partial t \partial z} + u\dfrac{\partial u}{\partial z} + v\dfrac{\partial v}{\partial z} + w\dfrac{\partial w}{\partial z} = -\dfrac{\partial F}{\partial z} - \dfrac{1}{\rho}\dfrac{\partial P}{\partial z} \end{cases} ---- (5)$$

Rewritten the system of equations given by equ (5), we have

$$-\frac{\partial}{\partial x}\left(\frac{\partial \phi}{\partial t}\right) + \frac{1}{2}\frac{\partial}{\partial x}(u^2 + v^2 + w^2) = -\frac{\partial F}{\partial x} - \frac{1}{\rho}\frac{\partial P}{\partial x} ---- (6)$$

$$-\frac{\partial}{\partial y}\left(\frac{\partial \phi}{\partial t}\right) + \frac{1}{2}\frac{\partial}{\partial y}(u^2 + v^2 + w^2) = -\frac{\partial F}{\partial y} - \frac{1}{\rho}\frac{\partial P}{\partial y} ----(7)$$

$$-\frac{\partial}{\partial z}\left(\frac{\partial \phi}{\partial t}\right) + \frac{1}{2}\frac{\partial}{\partial z}(u^2 + v^2 + w^2) = -\frac{\partial F}{\partial z} - \frac{1}{\rho}\frac{\partial P}{\partial z} ---- (8)$$

Recall few relations as:

$$d\left(\frac{\partial \phi}{\partial t}\right) = \frac{\partial}{\partial x}\left(\frac{\partial \phi}{\partial t}\right)dx + \frac{\partial}{\partial y}\left(\frac{\partial \phi}{\partial t}\right)dy + \frac{\partial}{\partial}\left(\frac{\partial \phi}{\partial t}\right)dz$$

$$dV = \frac{\partial F}{\partial x}dx + \frac{\partial F}{\partial y}dy + \frac{\partial F}{\partial z}dz$$

$$dP = \frac{\partial P}{\partial x}dx + \frac{\partial P}{\partial y}dy + \frac{\partial P}{\partial z}dz$$

$$d(u^2 + v^2 + w^2) = \frac{\partial}{\partial x}(u^2 + v^2 + w^2)dx$$

$$+ \frac{\partial}{\partial y}(u^2 + v^2 + w^2)dy + \frac{\partial}{\partial z}(u^2 + v^2 + w^2)dz$$

Multiplying (6), (7), and (8) by $dx, dy,$ and dz respectively, then add all the terms, we get

$$-\left[\frac{\partial}{\partial x}\left(\frac{\partial \phi}{\partial t}\right)dx + \frac{\partial}{\partial y}\left(\frac{\partial \phi}{\partial t}\right)dy + \frac{\partial}{\partial}\left(\frac{\partial \phi}{\partial t}\right)dz\right]$$

$$+ \frac{1}{2}\left[\frac{\partial}{\partial x}(u^2 + v^2 + w^2)dx + \frac{\partial}{\partial y}(u^2 + v^2 + w^2)dy\right]$$

$$+ \frac{1}{2}\left[\frac{\partial}{\partial z}(u^2 + v^2 + w^2)dz\right]$$

$$= -\left[\frac{\partial F}{\partial x}dx + \frac{\partial F}{\partial y}dy + \frac{\partial F}{\partial z}dz\right] + \frac{1}{\rho}\left[\frac{\partial P}{\partial x}dx + \frac{\partial P}{\partial y}dy + \frac{\partial P}{\partial z}dz\right]$$

On using above relations, we can rewritten the above equation as:

$$\Rightarrow -d\left(\frac{\partial \phi}{\partial t}\right) + \frac{1}{2}d(u^2 + v^2 + w^2) = -dF - \frac{1}{\rho}dP$$

$$\Rightarrow -d\left(\frac{\partial \phi}{\partial t}\right) + \frac{1}{2}d(u^2 + v^2 + w^2) + dF + \frac{1}{\rho}dP = 0$$

$$\Rightarrow -d\left(\frac{\partial \phi}{\partial t}\right) + \frac{1}{2}dV^2 + dF + \frac{1}{\rho}dP = 0 ---- (9)$$

Where $V^2 = u^2 + v^2 + w^2$ be the fluid velocity.

Here we consider ρ is the function of P [$i.e., \rho = f(P)$].
On integration the equation (9), we obtain

$$\boxed{-\frac{\partial \phi}{\partial t} + \frac{1}{2}V^2 + F + \int \frac{dP}{\rho} = K(constant) ---- (A)}$$

Equation (A) is known as **Bernoulli's equation** or **Pressure equation.**

DEDUCTION 1. The fluid be homogeneous *i.e.,* $\rho = constant$ and fluid is incompressible. In this case Bernoulli's equation applicable as

$$-\frac{\partial \phi}{\partial t} + \frac{1}{2} V^2 + F + \frac{P}{\rho} = k(constant)$$

DEDUCTION 2. The fluid motion be steady *i.e.*, $\frac{\partial \phi}{\partial t} = 0$. In this case Bernoulli's equation for an irrotational and incompressible fluids modified as

$$\frac{1}{2} V^2 + \frac{P}{\rho} = K(constant)$$

DEDUCTION 3. When we consider F be the gravitational force per unit mass and h be the vertical distance travelled by the fluid particle. Then we have,

$$F = -gh$$

In this case, Bernoulli's equation are applicable as

$$\frac{1}{2}V^2 + gh + \frac{P}{\rho} = K(constant)$$

BERNOULLI'S EQUATION FOR STREAMLINE

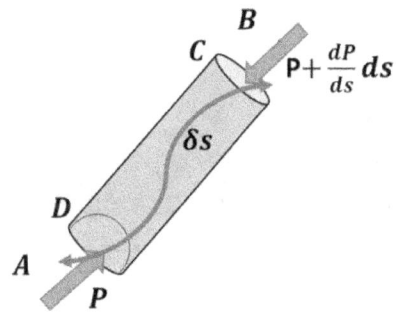

Consider a streamline AB in the fluid. Let δs be an element of this streamline and CD be a small cylindrical cross section area α and δs be the axis. Let v be the fluid motion and E be the component of external force per unit mass in the direction of streamline.

According to the Newton's second law of motion,

$$Mass \times acceleration = Total\ forces$$

$$\Rightarrow density \times volume \times \frac{Dv}{Dt} = External\ forces + Internal\ forces$$

$$\Rightarrow \rho.(\alpha\ \delta s) \times \frac{Dv}{Dt} = (\rho.\alpha\delta s)E + \left[P.\alpha - \left(P + \frac{dP}{ds}\delta s\right).\alpha\right]$$

[$E = external\ force\ per\ unit\ mass$

$\Rightarrow E.mass = external\ force\ on\ the\ fluid\ motion$

$\Rightarrow E.(density \times volume) = external\ force$

$\Rightarrow \rho.(cross\ section\ area \times height).E = external\ force$

$\Rightarrow \rho.(\alpha\ \delta s).E = external\ force$

$as\ we\ know,\ F = pressure \times cross\ section\ area$

$\Rightarrow Internal\ forces = change\ in\ pressure \times cross\ sectional\ area$]

$$\Rightarrow \rho.\alpha\,\delta s \times \frac{Dv}{Dt} = (\rho.\,\alpha\,\delta s)E + P.\alpha - P.\alpha - \frac{dP}{ds}\delta s.\alpha$$

$$\Rightarrow \rho.(\alpha\,\delta s) \times \frac{Dv}{Dt} = (\alpha\,\delta s)\left(\rho\,E - \frac{dP}{ds}\right)$$

$$\Rightarrow \rho\,\frac{Dv}{Dt} = \rho E - \frac{dP}{ds}$$

$$\Rightarrow \frac{Dv}{Dt} = E - \frac{1}{\rho}\frac{dP}{ds}$$

$$\Rightarrow \frac{dv}{dt} + v\frac{dv}{ds} = E - \frac{1}{\rho}\frac{dP}{ds} - - - (1)$$

$$\left[as\ material\ derivative\ for\ v:\ \frac{Dv}{Dt} = \frac{dv}{dt} + v\frac{dv}{ds}\right]$$

Here we consider the fluid motion is steady $i.e.,\frac{dv}{dt} = 0$

And the external force have potential function F such that

$$E = -\frac{dF}{ds}$$

Now equation (1) can be rewritten as

$$v\frac{dv}{ds} = -\frac{dF}{ds} - \frac{1}{\rho}\frac{dP}{ds}$$

$$\Rightarrow \frac{1}{2}\frac{dv^2}{ds} + \frac{dF}{ds} + \frac{1}{\rho}\frac{dP}{ds} = 0 - - - - (2)$$

Here we consider ρ is the function of P [$i.e., \rho = f(P)$].
On integration of equation (2) along the entire streamline ds, we obtain the Bernoulli's equation along the streamline as:

$$\frac{1}{2}v^2 + F + \int \frac{P}{\rho} = K$$

Where K be the constant of integration whose values depends on the streamlines.

For steady, incompressible flow:

$$\frac{P}{\rho} + \frac{v^2}{2} + F = K(constant)$$

This is the famous **Bernoulli's Equation** which is commonly used in fluid mechanics for steady, incompressible flow along a streamline in inviscid regions of flow.

DEDUCTIONS
DEDUCTION 1.

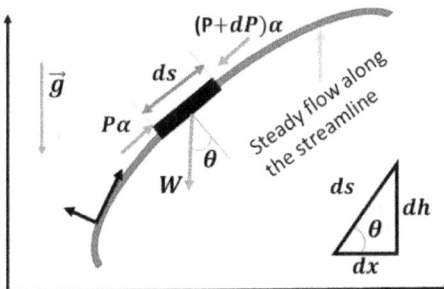

Figure showing the forces acting on a fluid particle along a streamline

If θ be the angle between the normal of the streamline and the vertical $y - axis$ at that point,

$$weight\ of\ the\ fluid = W$$
$$= mg = \rho\ (volume)g$$
$$\Rightarrow W = \rho g\ \alpha\ ds$$
$$\therefore\ External\ force$$
$$= \rho\ (\alpha\ ds)E$$
$$= -W \sin \theta$$
$$= \rho g \alpha\ ds \frac{dh}{ds}$$

$$\Rightarrow E = g \frac{dh}{ds}$$

Here, equation (1) can be rewritten as

$$\frac{dv}{dt} + v \frac{dv}{ds} = -W \sin \theta - \frac{1}{\rho} \frac{dP}{ds}$$

$$\Rightarrow \frac{dv}{dt} + \frac{1}{2} \frac{dv^2}{ds} = -g \frac{dh}{ds} - \frac{1}{\rho} \frac{dP}{ds} - -(3)$$

Consider the motion of a fluid particle in a flow field be steady and incompressible.

Then we have,

$$\frac{dv}{dt} = 0,\ and\ \rho = constant$$

$$(3) \Rightarrow \frac{1}{2} \frac{dv^2}{ds} = -g \frac{dh}{ds} - \frac{1}{\rho} \frac{dP}{ds}$$

$$\Rightarrow \frac{1}{\rho} \frac{dP}{ds} + \frac{1}{2} \frac{dv^2}{ds} + g \frac{dh}{ds} = 0 - - - - (4)$$

On integration of equation (4) along the entire streamline ds, we obtain the Bernoulli's equation along the streamline as:

$$\frac{P}{\rho} + \frac{1}{2} v^2 + gh = constant$$

DEDUCTION 2. The value of constant of integration in above equation can be evaluated at any point on the streamline when the pressure, density, velocity, and elevation are mentioned. The Bernoulli's equation for steady, incompressible flow can also be written between any two points on the same streamline as:

$$\frac{P_1}{\rho} + \frac{V_1^2}{2} + gh_1 = \frac{P_2}{\rho} + \frac{V_2^2}{2} + gh_2$$

DEDUCTION 3. If the fluid is flowing through a horizontal pipe, then h is constant. Then Bernoulli's equation becomes

$$\frac{P}{\rho} + \frac{v^2}{2} = k \ (constant)$$

$$\Rightarrow P \propto \frac{k}{v} \Rightarrow Pressure \propto \frac{1}{velocity}$$

Thus, Pressure P would be increases when the fluid motion v decreases in the case for the horizontal streamline flow.

DEDUCTION 4. <u>For unsteady, compressible flow:</u>
Mathematically, we take the velocity v of a fluid particle to be the function of s and t.
Consider the equation (1) as

$$\frac{dv}{dt} + v\frac{dv}{ds} = E - \frac{1}{\rho}\frac{dP}{ds}$$

$$\Rightarrow \frac{dv}{dt} + \frac{1}{2}\frac{dv^2}{ds} = -\frac{dF}{ds} - \frac{1}{\rho}\frac{dP}{ds}$$

$$\Rightarrow \frac{dv}{dt} + \frac{1}{2}\frac{dv^2}{ds} + \frac{dF}{ds} + \frac{1}{\rho}\frac{dP}{ds} = 0 ----(5)$$

Here, we assume ρ is the function of P [i.e., $\rho = f(P)$].
On integration of equation (5) along the entire streamline ds, we obtain

$$\int \frac{1}{\rho} dP + \int \frac{dv}{dt} + \frac{1}{2} v^2 + F = K(constant) --(6)$$

Consider the external force being the gravitational force per unit mass, so we can write

$$F = -gh$$

Then, the Bernoulli's equation given by equ (6) modified as:

$$\int \frac{1}{\rho} dP + \int \frac{dv}{dt} + \frac{1}{2} v^2 + gh = K(constant)$$

DEDUCTION 5. <u>MECHANICAL ENERGY BALANCE</u>
Recall the fact that energy can neither be created nor destroyed, however energy can only be converted from one form of energy to another. In other words, amount of the energy in working system remains same unless a force is applied to the system from a distance.

So that **Bernoulli's equation** can be expressed as *"the work done by the pressure and gravity forces on the fluids is equal to the increase in the kinetic energy of the fluid particles."*

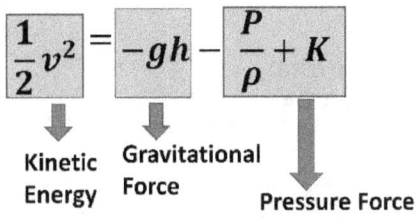

$$\frac{1}{2}v^2 = -gh - \frac{P}{\rho} + K$$

Kinetic Energy   Gravitational Force   Pressure Force

Similarly, Bernoulli's equation can be viewed as an expression of mechanical energy balance. We assumed that in regions of inviscid fluid, compressibility and frictional effects are ignored and the Bernoulli's equation is a mathematical expression where sum of the kinetic energy, potential energy, and the pressure energy of a fluid particle along the streamline remains constant for the steady flow through a streamline.

Therefore, **Bernoulli's equation** known as the "conservation of mechanical energy."

$$\boxed{\frac{P}{\rho}} + \boxed{\frac{v^2}{2}} + \boxed{gh} = K(constant)$$

Energy due to Pressure   Kinetic energy   Potential energy

DEDUCTION 6. Consider the Bernoulli's equation as:

$$\frac{P}{\rho} + \frac{1}{2}\,v^2 + gh = K \; (constant)$$

$$\Rightarrow P + \frac{1}{2}\rho v^2 + \rho gh = K'(constant) \; along \; the \; streamline$$

Here, $P =$ **static pressure** or the pressure used in thermodynamic system. Basically, it is the force per unit area that is exerted by a fluid upon a surface at rest. In other words, it is the resistance of a heating and cooling system's components to push airflows throughout the system. For example, the pressure inside the balloon.

$$\boxed{P} + \boxed{\frac{\rho v^2}{2}} + \boxed{\rho g h} = K(constant)$$

Static   Dynamic Hydrostatic
Pressure Pressure Pressure

Then, $\frac{\rho v^2}{2} = $ **dynamic pressure** which refer to the excess of pressure on the surface at which fluid is flowing brought to rest. For example, the dynamic pressure on the rocket launch increases because of the increasing the velocity from zero (initial) to some maximum value (Max Q).

Note: The sum of the free-stream static pressure and the free-stream dynamic pressure is equal to stagnation pressure. Stagnation pressure is the pressure the fluid would experiences by bringing the fluid to rest without loss of mechanical energy.

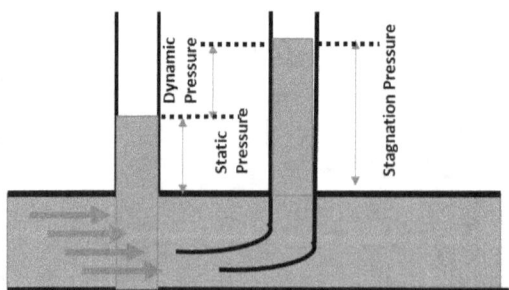

Next, $\rho g h = $ **Hydrostatic pressure** which refers as the pressure exerted by a fluid on the surface of flow at any point of time due to the force of gravity. Its value depends on the selection of reference level. For example, Fluid experience some pressure from the wall of the container when fluid is filled in closed container.

BERNOULLI'S EQUATION FOR COMPRESSIBLE FLOW

Consider the fluid is an ideal gas. The ideal gas flow is steady where viscous effects and frictional losses are negligible.

However, compressibility effects in fluid flowing cannot be ignored. Then, consider the Bernoulli's equation as:

$$\frac{1}{2}v^2 + F + \int \frac{P}{\rho} = K \ (constant) \ \ along \ a \ streamline --(1)$$

Case 1. <u>Isothermal condition</u>:

Here, we assume that temperature T of a system remains constant. The expansion of air under constant temperature refers to thermal expansion. The air does work on expanding, it loses heat subsequently, heat must be added to the air to maintain it at constant temperature. So, we have relation for isothermal process as:

$$P = \rho RT$$

$$\Rightarrow \rho = \frac{P}{RT}$$

$$\Rightarrow \int \frac{P}{\rho} = \int \frac{P}{P/RT} = RT \int \frac{dP}{P} = RT \ \ln P$$

Then, equ (1) becomes

$$\frac{1}{2}v^2 + F + RT \ \ln P \ = K \ (constant)$$

Case 2. <u>Isentropic condition</u>:

Here, we assume the thermodynamic system is both adiabatic and reversible. Isentropic process is a thermodynamic process in which the entropy of any fluid remains fixed. Moreover, change in volume results in a change in pressure and temperature of the ideal gas which evaluate the amount of work the nozzle can diffuse. So, we have the relation for isentropic process as:

$$\frac{P}{\rho^\gamma} = C \ (constant)$$

Where $\gamma = \frac{C_p}{C_v} = ratio \ of \ specific \ heats$

$$\Rightarrow \rho = C^{-\frac{1}{\gamma}} P^{\frac{1}{\gamma}}$$

$$\Rightarrow \int \frac{dP}{\rho} = \int C^{\frac{1}{\gamma}} P^{-\frac{1}{\gamma}} \, dP = C^{\frac{1}{\gamma}} \left(\frac{P^{-\frac{1}{\gamma}+1}}{-\frac{1}{\gamma}+1} \right)$$

$$\Rightarrow \int \frac{dP}{\rho} = \left(\frac{P^{\frac{1}{\gamma}}}{\rho} \right) \left(\frac{P^{-\frac{1}{\gamma}+1}}{-\frac{1}{\gamma}+1} \right) = \left(\frac{\gamma}{\gamma-1} \right) \left(\frac{P}{\rho} \right)$$

On substitute above relation on the Bernoulli's equation, we obtain as:

$$\left(\frac{\gamma}{\gamma-1}\right)\left(\frac{P}{\rho}\right)+\frac{1}{2}v^2+F=K\ (constant)$$

DEDUCTION: consider an isentropic process where a gas is diffuses from the rest with negligible change in elevation.

$$\therefore h_1=h_2\ \ and\ \ v_1=0$$

Now apply the Bernoulli equation at the two points in a streamline, we obtain

$$\left(\frac{\gamma}{\gamma-1}\right)\left(\frac{P_1}{\rho_1}\right)+\frac{1}{2}\overset{0}{\cancel{v_1^2}}+g\cancel{h_1}=\left(\frac{\gamma}{\gamma-1}\right)\left(\frac{P_2}{\rho_2}\right)+\frac{1}{2}v_2^2+g\cancel{h_2}$$

$$\Rightarrow\left(\frac{\gamma}{\gamma-1}\right)\left(\frac{P_1}{\rho_1}\right)=\left(\frac{\gamma}{\gamma-1}\right)\left(\frac{P_2}{\rho_2}\right)+\frac{1}{2}v_2^2\ ---(2)$$

Now consider the Mach number defined as $Ma=v/c$ where $c=\sqrt{kRT}$ i.e., c is the local speed of sound for ideal gas.

Then equ (2) simplifies as

$$\frac{P_1}{P_2}=\left[1+\left(\frac{\gamma-1}{2}\right)Ma_2^2\right]^{\frac{\gamma}{\gamma-1}}$$

RESTRICTIONS ON ITS APPLICABILITY

The Bernoulli's equation is one of the most widely used equation in fluid dynamics. However, the attributes are easy to used but sometime attempt to use in inappropriate manner. Thus, it is important to discuss the major restrictions on its applicability and analyzed the limitations as mentioned below:

1. STEADY FLOW: The most important restriction on the applicability of Bernoulli's equation is that fluid motion would be steady. If we assume that the fluid motion is unsteady then, various parameters changes the condition during the transient period of motion. Thus, for simplicity in calculations of elevation and pressure for the respective fluid motion, we assumed the steady flow.

2. INCOMPRESSIBLE FLOW: Next, very important restrictions on the applicability of Bernoulli's equation taken as the fluid motion is an incompressible flow such that $\rho=constant$. This condition is satisfied by liquid as well as gases at Mach number less than about 0.3 because at such low velocity, compressibility variations and density variations can be ignored for calculations purpose.

3. Negligible Frictional effects: In real life, every fluid has some amount of viscosity or we can say there has some number of frictional forces for flowing of fluids. Also, it has been observed that some frictional forces close to the solid surfaces. However, Bernoulli's equation is applicable along the streamline in flowing region of fluids instead of along a streamline very close to the surface.

4. Fluid particle moves along a streamline: Numerically, we used Bernoulli's equation along a streamline so that the constant *K* values would be different for the different directional streamlines. Considering fluid motion is irrotational such that there is no vorticity in the fluid motion field. Thus, we are concern about the streamlines on its application and also, we can apply the equation between two points in the irrotational region of the fluid flow.

5. Ignorance of shaft work: Bernoulli's equation cannot applicable for the flow in pumps, turbine, fan, and any impeller because devices carry out energy interactions with the fluid particles which disrupt the streamlines.

6. Omission of temperature variations: we all know that the density of a gas is inversely proportional to temperature. So, significant variation in temperature in heating and cooling system should be ignored while apply the Bernoulli's equation to obtain numerical values.

3.APPLICANTS OF THE EQUATION

DETERMINATION OF FLOW RATE THROUGH PIPE

Flow rate through the pipe is calculate by recording the pressure drop across the coaxial area contraction. Thus, determination of flow rate from the reading of pressure drop depends on the application of Bernoulli's equation. There are three different types of flow meters such as

1. Venturimeter
2. Orificemeter
3. Pitot tube
4. Nozzle

VENTURIMETER

Construction: A venturimeter comprise of a pipe having two conical parts with a uniform cross-section in between. This short portion known as throat which has the minimum area. The two conical parts have almost same diameters, but one is of shorter length with larger angle while the other is of longer length with small angle.

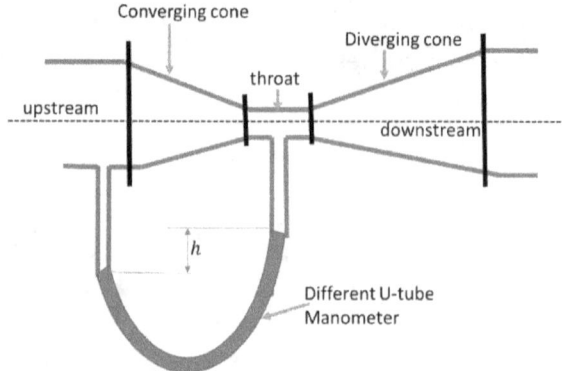

Working: The fluid flow inside venturimeter in such a way that upstream flow takes place through the short conical portion while the downstream flow through the larger conical portion. This ensures a rapid converging passage and a gradual diverging passage in the direction of flow to avoid the loss of energy due to separation. For inflow through converging part, the velocity increases according to the principle of continuity, while the pressure decreases according to the Bernoulli's principle. The fluid's velocity will be maximum and pressure reaches its minimum value at the throat point. Next, decrease in velocity and increase in pressure takes place in outflow through diverging part.

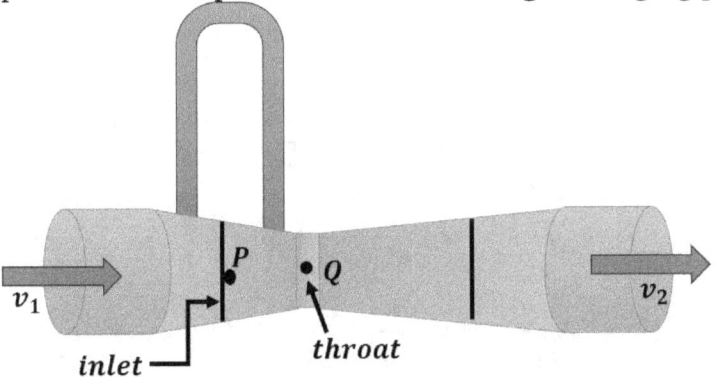

Consider the fluid inflow through converging part of cross section area A_1 to the throat of cross section area A_2. Also, Let v_1 and v_2 be the fluid velocities and P_1 and P_2 are the pressure at two points P and Q. Moreover, a differential manometeric U-tube is fitted with venturimeter as shown in figure.

According to the continuity principle, we have

$$v_1 A_1 = v_2 A_2$$
$$\Rightarrow v_1 = v_2 \frac{A_2}{A_1}$$

Next, applying the Bernoulli's equation along the streamline from the point P to the point Q, we obtain as:

$$\frac{P_1}{\rho g} + \frac{v_1^2}{2g} + h_1 = \frac{P_2}{\rho g} + \frac{v_2^2}{2g} + h_2$$

Since venturimeter is placed horizontally, so

$$h_1 = h_2$$
$$\Rightarrow \frac{P_1}{\rho g} + \frac{v_1^2}{2g} = \frac{P_2}{\rho g} + \frac{v_2^2}{2g}$$
$$\Rightarrow \frac{1}{2g} v_2^2 = \frac{1}{\rho g} (P_1 - P_2) + \frac{1}{2g} v_2^2 \left(\frac{A_2}{A_1}\right)^2$$

$$\Rightarrow \frac{1}{2g}v_2^2\left[1-\left(\frac{A_2}{A_1}\right)^2\right] = \frac{P_1 - P_2}{\rho g} \quad ----(1)$$

The pressure difference between inlet and throat sections is defined as $P_1 - P_2 = \rho g h$

$$\Rightarrow \frac{P_1 - P_2}{\rho g} = h$$

Then equ (1) becomes $\frac{1}{2g}v_2^2\left[1-\left(\frac{A_2}{A_1}\right)^2\right] = h$

$$\Rightarrow h = \frac{v_2^2}{2g}\left[\frac{A_1^2 - A_2^2}{A_1^2}\right] \Rightarrow v_2^2 = 2gh\frac{A_1^2}{A_1^2 - A_2^2}$$

$$\Rightarrow v_2 = \sqrt{2gh\frac{A_1^2}{A_1^2 - A_2^2}}$$

$$\Rightarrow v_2 = \frac{A_1}{\sqrt{A_1^2 - A_2^2}}\sqrt{2gh} \quad ----(2)$$

Let Q be the volumetric discharge of the fluid through the pipe which is defined as

$$Q = v_2 A_2$$

$$\Rightarrow Q = \frac{A_1}{\sqrt{A_1^2 - A_2^2}}\sqrt{2gh}.A_2$$

$$\Rightarrow Q = \frac{A_1 A_2}{\sqrt{A_1^2 - A_2^2}}\sqrt{2gh} \quad -----(3)$$

Equation (3) provides the value of theoretical discharge under ideal conditions. Actual discharge will be less than theoretical discharge and can be calculated as

$$Q_{actual} = C_d \times \frac{A_1 A_2}{\sqrt{A_1^2 - A_2^2}}\sqrt{2gh} \quad -----(4)$$

Where C_d is the coefficient of discharge on using venturimeter and its value is less than 1.

Possible value of h indicates by Differential U-tube manometer

 ❖ When differential manometer filled with a fluid which is heavier than a fluid flowing through pipe. Let

ρ_h = density of the heavier fluid

ρ_P = density of the fluid flows through pipe

y = difference of the heavier fluid column in U-tube

Then value of h determine from

$$h = y \left[\frac{\rho_h}{\rho_p} - 1 \right]$$

❖ When the differential manometer filled with a fluid which is lighter than the fluid flows through the pipe. Then value of h determine from

$$h = y \left[1 - \frac{\rho_h}{\rho_p} \right] \text{ where}$$

ρ_h = density of the heavier fluid

ρ_p = density of the fluid flows through pipe

y = difference of the heavier fluid column in U-tube

DEDUCTION:

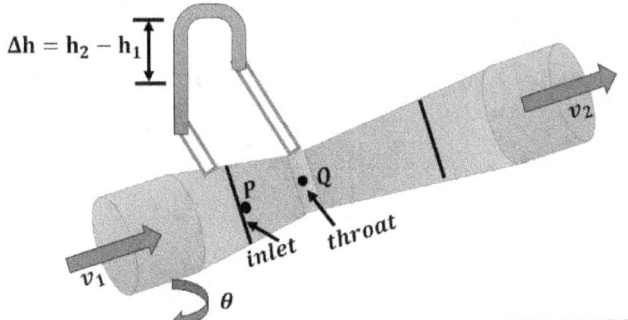

Consider a venturimeter positioned at an inclined line at a certain angle to the horizon to measure the flow rate through the pipe as shown in figure. Let us consider a steady, incompressible fluid such that the velocity and pressure at any section will be uniform. Then, the velocity and pressure at the converging part (point P) and at throat (point Q) are v_1, P_1 and v_2, P_2. Here the vertical column heights at the point P and point Q be h_1 and h_2 ($h_2 > h_1$) so that assume that $\Delta h = h_2 - h_1$.

Next, applying the Bernoulli's equation along the streamline from the point P to the point Q, we obtain as:

$$\frac{P_1}{\rho g} + \frac{v_1^2}{2g} + h_1 = \frac{P_2}{\rho g} + \frac{v_2^2}{2g} + h_2$$

$$\Rightarrow \frac{1}{2}(v_2^2 - v_1^2) = \frac{1}{\rho}(P_1 - P_2) + g(h_1 - h_2)$$

$$\Rightarrow \frac{1}{2}(v_2^2 - v_1^2) = \frac{1}{\rho}(P_1 - P_2) - g(h_2 - h_1) - - - -(3)$$

According to the continuity principle, we have

$$v_1 A_1 = v_2 A_2 \Rightarrow v_1 = v_2 \frac{A_2}{A_1}$$

Now, the equ (3) becomes

$$\Rightarrow \frac{1}{2}\left[v_2^2 - v_2^2\left(\frac{A_2}{A_1}\right)^2\right] = \frac{1}{\rho}(P_1 - P_2) - g(h_2 - h_1)$$

$$\Rightarrow \frac{1}{2}v_2^2\left[1 - \left(\frac{A_2}{A_1}\right)^2\right] = \frac{1}{\rho}(P_1 - P_2) - g(h_2 - h_1)$$

$$\Rightarrow v_2 = \sqrt{\frac{2\left[\frac{(P_1 - P_2)}{\rho} - g(h_2 - h_1)\right]}{\left[1 - \left(\frac{A_2}{A_1}\right)^2\right]}}$$

Thus, Q_2 be the volumetric discharge of the fluid through the pipe which is defined as

$$Q_2 = v_2 A_2$$

$$\Rightarrow Q_2 = A_2 \sqrt{\frac{2\left[\frac{(P_1 - P_2)}{\rho} - g(h_2 - h_1)\right]}{\left[1 - \left(\frac{A_2}{A_1}\right)^2\right]}}$$

$$\Rightarrow Q_2 = \frac{A_2}{\sqrt{\left[1 - \left(\frac{A_2}{A_1}\right)^2\right]}} \sqrt{2\left[\frac{(P_1 - P_2)}{\rho} - g(h_2 - h_1)\right]} - - -(4)$$

When the difference of pressure between two points P and Q by using U-tube manometer as shown in figure, we have

$$P_1 + \rho g (h_1 - h_o) = P_2 + \rho g(h_2 - h_o - \Delta h) + \Delta h\, \rho_m g$$

$$\Rightarrow (P_1 + \rho g\, h_1) - (P_2 + \rho g\, h_2) = (\rho_m - \rho)g\, \Delta h$$

$$\Rightarrow \left[\frac{(P_1 - P_2)}{\rho} - g(h_2 - h_1)\right] = \left[\frac{\rho_m}{\rho} - 1\right]\Delta h$$

Thus, equ (4) becomes

$$Q_2 = \frac{A_1 A_2}{\sqrt{A_1^2 - A_2^2}}\sqrt{2g\left[\frac{\rho_m}{\rho} - 1\right]\Delta h} \, ---- (5)$$

The expression of volumetric flow rate given by equ (5), in terms of manometer deflection Δh, suitable for all conditions like as whether venturimeter pipe placed horizontal or at some inclined angle. Usually, the measured value of Δh will always greater than that of ideal fluid due to the frictional losses during the fluid flow through the pipe. In order to get actual value of flow rate, an multiplying factor C_d known as coefficient of discharge is introduce in the equ (5) as

$$Q_2 = C_d \frac{A_1 A_2}{\sqrt{A_1^2 - A_2^2}}\sqrt{2g\left[\frac{\rho_m}{\rho} - 1\right]\Delta h}$$

Where the coefficient of discharge C_d is defined as

$$C_d = \frac{actual\,rate\,of\,discharge}{Theoretical\,rate\,of\,discharge} < 1$$

Note: The value of C_d for a venturimeter usually lies between 0.95 to 0.98.

Application: Venturimeter is a device used to measure the flow rate of the steady incompressible fluid. It is commonly used in water supply industry, HVAC systems, Power generation systems, water treatment plant etc. Few major applications as mention below:
1. It is used in Engine Carburetors to measure airflow relevant in automobile industry.
2. It is commonly used in process industries to control flow rate.
3. It is also used in medical industry as blood flows in the arteries measured by this.
4. The flow rate of oil and gasoline taken by venturimeters in Oil & Gas Industries.

Advantages: The venturimeter is widely used in many industries because of following advantages as
1. It commonly provide accurate reading.

2. The system is not dependent on temperature and pressure inside the pipe.
3. Throughout the fluid flow, there has been a very low head loss of mechanical energy.
4. Different types of fluids including the suspension of solids, gases, slurries, chemicals, dirty liquids with basic water can be used for its working.
5. There is very less chance of flow clogged inside the pipe.
6. The system can be operate with high discharge coefficient and with very low pressure drop in the pipe.
7. The system is well adapted like as it can installed as horizontal, inclined, and vertical direction also.

Disadvantages: However, there are certain limitations such as:

1. This system is not suitable for flow through very small diameter pipes.
2. The system is very expensive as compared to Orificemeter.
3. Venturimeter is large in dimension so it occupies large area for installation.

SOLVED EXAMPLES

EXAMPLE 1. Determine the discharge rate of flow of water through a horizontal ventrurimeter with inlet and throat diameters 40 cm and 20 cm respectively. Also, the reading of differential manometer connected to the inlet and the throat is 20 cm of mercury. It is given that the coefficient of discharge (C_d) is 0.98

Ans: In this case,

Let d_1 = diameter of inlet = 40 cm

A_1 = cross-sectional area of inlet = $\dfrac{\pi}{4} d_1^2 = \dfrac{\pi}{4}(40)^2 = 1256$ cm^2

d_2 = diameter of throat = 20 cm

A_2 = cross-sectional area of throat = $\dfrac{\pi}{4} d_2^2 = \dfrac{\pi}{4}(20)^2 = 314$ cm^2

C_d = co-efficient of discharge = 0.98

ρ_m = density of mercury in manometer = 13545.848 kg/m^3

ρ_w = density of water flows through pipe = 1000 kg/m^3

y = difference of the mercury level in manometric U-tube = 20 cm

The difference of pressure head is given by

$$h = y\left[\frac{\rho_m}{\rho_w} - 1\right] = 20\left[\frac{13545.848}{1000} - 1\right]$$

$$\Rightarrow h = 20[12.546] = 250.92 \text{ cm of water}$$

The discharge through ventrurimeter is

$$Q = C_d \frac{A_1 A_2}{\sqrt{A_1^2 - A_2^2}}\sqrt{2gh}$$

$$\Rightarrow Q = 0.98 \times \frac{1256 \times 314}{\sqrt{(1256)^2 - (314)^2}}\sqrt{2 \times 9.8 \times 250.92}$$

$$\Rightarrow Q = 0.98 \times \frac{1256 \times 314}{\sqrt{1478940}}\sqrt{4918.032}$$

$$\Rightarrow Q = 0.98 \times \frac{394384}{1216.12} \times 70.12$$

$$\Rightarrow Q = 0.98 \times 324.3 \times 70.12$$

$$\Rightarrow Q = 22285 \text{ cm}^3/\text{sec} = \frac{22285}{1000} \text{ Lt/sec}$$

$$\Rightarrow Q = 22.285 \text{ Lt/sec}$$

Example 2. The discharge of oil of density 700 kg/m^3 through horizontal ventrurimeter is 50 Lt/sec. The diameters of inlet and throat of the horizontal ventrurimeter are 30 cm and 15 cm respectively which used to measure the flow of oil. Determine the oil-mercury differential manometric reading provided co-efficient of discharge (C_d) is 0.98.

Ans: In this case,

d_1 = diameter of inlet = 30 cm

A_1 = cross-sectional area of inlet $= \dfrac{\pi}{4} d_1^2 = \dfrac{\pi}{4}(30)^2 = 706.5 \text{ cm}^2$

d_2 = diameter of throat section = 15 cm

A_2 = cross-sectional area of throat $= \dfrac{\pi}{4} d_2^2 = \dfrac{\pi}{4}(15)^2 = 176.625 \text{ cm}^2$

Q = discharge of oil = 50 Lt/sec $= 5 \times 10^4 \text{ cm}^3 / \text{sec}$

C_d = co-efficient of discharge $= 0.98$

ρ_m = density of mercury $= 13545.848 \ kg / m^3$

ρ_o = density of oil $= 700 \ kg / m^3$

Let y be difference of the mercury level in manometric U-tube

The discharge through ventrurimeter is

$$Q = C_d \frac{A_1 A_2}{\sqrt{A_1^2 - A_2^2}} \sqrt{2gh}$$

$$\Rightarrow 5 \times 10^4 = 0.98 \times \frac{706.5 \times 176.625}{\sqrt{(706.5)^2 - (176.625)^2}} \sqrt{2gh}$$

$$\Rightarrow 5 \times 10^4 = 0.98 \times \frac{706.5 \times 176.625}{684.1} \sqrt{2gh}$$

$$\Rightarrow \sqrt{2gh} = \frac{5 \times 10^4 \times 684.1}{0.98 \times 706.5 \times 176.625}$$

$$\Rightarrow \sqrt{2gh} = \frac{34205000}{122289.85}$$

$$\Rightarrow \sqrt{2gh} \ ; \ 280$$

$$\Rightarrow 2gh = 78400$$

$$\Rightarrow h = \frac{78400}{2g} = \frac{78400}{2 \times 9.8} = \frac{78400}{19.6}$$

$$\Rightarrow h = 4000$$

$$\Rightarrow y\left[\frac{\rho_m}{\rho_o} - 1\right] = 4000$$

$$\Rightarrow y\left[\frac{13545.848}{700} - 1\right] = 4000$$

$$\Rightarrow y[18.35] = 4000$$

$$\Rightarrow y = \frac{4000}{18.35}$$

$$\Rightarrow y = 218 \text{ cm of mercury}$$

EXAMPLE 3. The diameters of inlet and throat of a horizontal ventrurimeter is 30 cm and 15 cm. the pressure at inlet is 18 N/cm^2 and the vacuum pressure at the throat is 30 cm of mercury. Determine the discharge of water through ventrurimeter provided the co-efficient of discharge (C_d) is 0.98.

Ans: In this case,

d_1 = diameter of inlet = 30 cm

A_1 = cross-sectional area of inlet = $\frac{\pi}{4}d_1^2 = \frac{\pi}{4}(30)^2 = 706.5 \text{ cm}^2$

d_2 = diameter of throat = 15 cm

A_2 = cross-sectional area of throat = $\frac{\pi}{4}d_2^2 = \frac{\pi}{4}(15)^2 = 176.625 \text{ cm}^2$

P_1 = pressure at inlet = 18 $N/cm^2 = 18 \times 10^4$ N/m^2

ρ = density of water = 1000 kg/m^3

$\therefore \quad \frac{P_1}{\rho g} = \frac{18 \times 10^4}{10^3 \times 9.81}$; 18 m of water

$$\frac{P_2}{\rho g} = -30 \text{ cm of mercury}$$

$$\Rightarrow \frac{P_2}{\rho g} = -0.30 \text{ m of mercury}$$

$$\Rightarrow \frac{P_2}{\rho g} = -0.30 \times 13.6 = -4.08 \text{ m of water}$$

∴   Difference of pressure heads $= h = \dfrac{P_1}{\rho g} - \dfrac{P_2}{\rho g}$

$$\Rightarrow h = 18 + 4.08 = 22.08 \text{ m of water}$$
$$\Rightarrow h = 2208 \text{ cm of water}$$

The discharge through ventrurimeter is

$$Q = C_d \frac{A_1 A_2}{\sqrt{A_1^2 - A_2^2}} \sqrt{2gh}$$

$$\Rightarrow Q = 0.98 \times \frac{706.5 \times 176.625}{\sqrt{(706.5)^2 - (176.625)^2}} \sqrt{2 \times 9.81 \times 2208}$$

$$\Rightarrow Q = 0.98 \times \frac{124785.5625}{684.1} \sqrt{43320.96}$$
$$\Rightarrow Q = 0.98 \times 182.4 \times 208.14$$
$$\Rightarrow Q = 37205 \text{ cm}^3 / \sec$$
$$\Rightarrow Q = 37.205 \text{ Lt/sec}$$

EXAMPLE 4. Determine the discharge of water flows through a pipe of diameter 40 cm positioned in an inclined manner and a ventrurimeter is inserted whose throat diameter is 20 cm. The differential manometer gives a reading of 30 cm of mercury. The loss of head between the inlet and throat is 0.3 times the kinetic head of the pipe.

Ans: In this case,

d_1 = diameter of inlet of the pipe = 40 cm

A_1 = cross-sectional area of inlet = $\dfrac{\pi}{4}d_1^2 = \dfrac{\pi}{4}(40)^2 = 1256$ cm^2

d_2 = diameter of throat = 20 cm

A_2 = cross-sectional area of throat = $\dfrac{\pi}{4}d_2^2 = \dfrac{\pi}{4}(20)^2 = 314$ cm^2

ρ_m = density of mercury in manometer = 13545.848 $kg\,/\,m^3$

ρ_w = density of water flows through pipe = 1000 $kg\,/\,m^3$

y = difference of the mercury level in manometric U-tube = 30 cm

Difference of pressure head is

$$h = y\left[\frac{\rho_m}{\rho_w} - 1\right]$$

$$\Rightarrow h = 30\left[\frac{13545.848}{1000} - 1\right]$$

$$\Rightarrow h = 30 \times 12.54$$

$$\Rightarrow h = 376.2 \text{ cm of water}$$

$\therefore \quad \left(\dfrac{P_1}{\rho g} + z_1\right) - \left(\dfrac{P_2}{\rho g} + z_2\right) = h = 376.2$

Loss of head, $h_L = 0.3 \times$ kinetic head of pipe $= 0.3 \times \dfrac{v_1^2}{2g}$

Now applying Bernoulli's equation along the streamline from the point P to point Q, we obtain as

$$\frac{P_1}{\rho g}+\frac{v_1^2}{2g}+z_1=\frac{P_2}{\rho g}+\frac{v_2^2}{2g}+z_2+h_L$$

$$\Rightarrow \left(\frac{P_1}{\rho g}+z_1\right)-\left(\frac{P_2}{\rho g}+z_2\right)+\frac{v_1^2}{2g}-\frac{v_2^2}{2g}=h_L$$

$$\Rightarrow h+\frac{v_1^2}{2g}-\frac{v_2^2}{2g}=h_L$$

$$\Rightarrow 376.2+\frac{v_1^2}{2g}-\frac{v_2^2}{2g}=0.3\times\frac{v_1^2}{2g}$$

$$\Rightarrow 376.2+0.7\times\frac{v_1^2}{2g}-\frac{v_2^2}{2g}=0 \qquad \text{-------(1)}$$

Now apply continuity equation along the streamline from the point P to point Q, we obtain as

$$A_1v_1=A_2v_2$$

$$\Rightarrow v_1=\frac{A_2}{A_1}v_2$$

$$\Rightarrow v_1=\frac{314}{1256}v_2$$

$$\Rightarrow v_1=\frac{v_2}{4}$$

Using above relation, equation (1) reduced to

$$376.2+\frac{0.7}{2g}\left(\frac{v_2}{4}\right)^2-\frac{v_2^2}{2g}=0$$

$$\Rightarrow 376.2+\frac{v_2^2}{2g}\left[\frac{0.7}{16}-1\right]=0$$

$$\Rightarrow \frac{v_2^2}{2g}(-0.96)=-376.2$$

$$\Rightarrow v_2^2=\frac{376.2\times2\times9.81}{0.96}; \ 7688$$

$$\Rightarrow v_2 = \sqrt{7688} \; ; \; 88 \text{ cm/sec}$$

Discharge through ventrurimeter is

$$Q = A_2 v_2$$

$$\Rightarrow Q = 314 \times 88$$
$$\Rightarrow Q = 27632 \text{ cm}^3 / \text{sec}$$
$$\Rightarrow Q = 27.632 \text{ Lt/sec}$$

EXAMPLE 5. A horizontal ventrurimeter with inlet and throat diameters 30 cm and 15 cm respectively is used to measure the flow of water. The differential manometer installed in between inlet and throat gives the reading 10 cm of mercury. Determine the discharge rate of flow provided the co-efficient of discharge (C_d) is 0.98.

Ans: In this case,

Let d_1 = diameter of inlet = 30 cm

A_1 = cross-sectional area of inlet = $\dfrac{\pi}{4} d_1^2 = \dfrac{\pi}{4}(30)^2 = 706.5 \text{ cm}^2$

d_2 = diameter of throat = 15 cm

A_2 = cross-sectional area of throat = $\dfrac{\pi}{4} d_2^2 = \dfrac{\pi}{4}(15)^2 = 176.625 \text{ cm}^2$

C_d = co-efficient of discharge = 0.98

ρ_m = density of mercury in manometer = 13545.848 kg / m^3

ρ_w = density of water flows through pipe = 1000 kg / m^3

y = difference of the mercury level in manometric U-tube = 10 cm

The difference of pressure head is given by

$$h = y\left[\frac{\rho_m}{\rho_w} - 1\right] = 10\left[\frac{13545.848}{1000} - 1\right]$$

$$\Rightarrow h = 10(12.545)$$

$$\Rightarrow h = 120.55 \text{ cm of water}$$

Discharge through ventrurimeter is

$$Q = C_d \frac{A_1 A_2}{\sqrt{A_1^2 - A_2^2}} \sqrt{2gh}$$

$$\Rightarrow Q = 0.98 \times \frac{706.5 \times 176.625}{\sqrt{(706.5)^2 - (176.625)^2}} \sqrt{2 \times 9.81 \times 120.55}$$

$$\Rightarrow Q = 0.98 \times 182.4 \times \sqrt{2365.191}$$

$$\Rightarrow Q = 0.98 \times 182.4 \times 48.63$$

$$\Rightarrow Q = 8692.71 \text{ cm}^3 / \text{sec}$$

$$\Rightarrow Q = 8.7 \text{ Lt/sec}$$

EXAMPLE 6. An oil of density 800 kg/m^3 is flowing through a ventrurimeter having inlet diameter 30 cm and throat diameter 15 cm. The oil-mercury differential manometer shows a reading of 15 cm. Determine the discharge of oil through the horizontal ventrurimeter provided the co-efficient of discharge (C_d) is 0.98.

Ans: In this case,

Let d_1 = diameter of inlet = 30 cm

A_1 = cross-sectional area of inlet = $\frac{\pi}{4}d_1^2 = \frac{\pi}{4}(30)^2 = 706.5 \text{ cm}^2$

d_2 = diameter of throat = 15 cm

A_2 = cross-sectional area of throat = $\frac{\pi}{4}d_2^2 = \frac{\pi}{4}(15)^2 = 176.625 \text{ cm}^2$

C_d = co-efficient of discharge = 0.98

ρ_m = density of mercury in manometer = 13545.848 kg/m^3

ρ_w = density of oil flows through pipe = 800 kg/m^3

y = difference of the mercury level in manometric U-tube = 15 cm

The difference of pressure head is given by

$$h = y\left[\frac{\rho_m}{\rho_w} - 1\right] = 15\left[\frac{13545.848}{800} - 1\right]$$
$$\Rightarrow h = 15(15.93)$$
$$\Rightarrow h = 238.95 \text{ cm of water}$$

Discharge through ventrurimeter is

$$Q = C_d \frac{A_1 A_2}{\sqrt{A_1^2 - A_2^2}}\sqrt{2gh}$$

$$\Rightarrow Q = 0.98 \times \frac{706.5 \times 176.625}{\sqrt{(706.5)^2 - (176.625)^2}}\sqrt{2 \times 9.81 \times 238.95}$$

$$\Rightarrow Q = 0.98 \times 182.4 \times \sqrt{4688}$$
$$\Rightarrow Q = 0.98 \times 182.4 \times 68.47$$
$$\Rightarrow Q = 12238 \text{ cm}^3/\text{sec}$$
$$\Rightarrow Q = 12.23 \text{ Lt/sec}$$

EXAMPLE 7. A horizontal ventrurimeter with inlet diameter 30 cm and throat diameter 15 cm is used to measure the rate of flow of oil of density 700 kg/m^3. the discharge of oil through ventrurimeter is 50 Lt/sec, Calculate the oil-mercury differential manometric reading provided the co-efficient of discharge (C_d) is 0.98.

Ans: In this case,

Let d_1 = diameter of inlet = 30 cm

A_1 = cross-sectional area of inlet = $\dfrac{\pi}{4}d_1^2 = \dfrac{\pi}{4}(30)^2 = 706.5\ \text{cm}^2$

d_2 = diameter of throat = 15 cm

A_2 = cross-sectional area of throat = $\dfrac{\pi}{4}d_2^2 = \dfrac{\pi}{4}(15)^2 = 176.625\ \text{cm}^2$

C_d = co-efficient of discharge = 0.98

ρ_m = density of mercury in manometer = 13545.848 kg/m^3

ρ_0 = density of oil flows through pipe = 700 kg/m^3

y = difference of the mercury level in manometric U-tube

Q = Discharge through ventrurimeter = 50 Lt/sec

$\Rightarrow Q = 5 \times 10^4\ cm^3/\sec$

However, discharge through ventrurimeter is defined as

$$Q = C_d \frac{A_1 A_2}{\sqrt{A_1^2 - A_2^2}}\sqrt{2gh}$$

$$\Rightarrow 5\times10^4 = 0.98 \times \frac{706.5 \times 176.625}{\sqrt{(706.5)^2 - (176.625)^2}}\sqrt{2gh}$$

$$\Rightarrow 5\times10^4 = 0.98 \times 182.4 \times \sqrt{2gh}$$

$$\Rightarrow \sqrt{2gh} = \frac{5\times10^4}{0.98\times182.4} = \frac{5\times10^4}{178.752} = 279.717$$

$$\Rightarrow 2gh = (279.717)^2 = 78241$$

$$\Rightarrow h = \frac{78241}{2\times9.81} = \frac{78241}{19.62}$$

$$\Rightarrow h\ ;\ 3988\ \text{cm of water}$$

Also, difference in pressure heads is defined as

$$h = y\left[\frac{\rho_m}{\rho_w} - 1\right] \Rightarrow 3988 = y\left[\frac{13545.848}{700} - 1\right]$$

$$\Rightarrow 3988 = y(18.35)$$

$$\Rightarrow y = \frac{3988}{18.35} = 217 \text{ cm}$$

EXAMPLE 8. A horizontal ventrurimeter with inlet diameter 30 cm and throat diameter 15 cm is used to measure the flow of water through pipe. The pressure at inlet is 15 N/cm^2 and vacuum pressure at the throat is 40 cm of mercury. Determine the discharge of water through ventrurimeter provided the co-efficient of discharge (C_d) is 0.98.

Ans: In this case,

d_1 = diameter of inlet = 30 cm

A_1 = cross-sectional area of inlet = $\frac{\pi}{4}d_1^2 = \frac{\pi}{4}(30)^2 = 706.5 \text{ cm}^2$

d_2 = diameter of throat = 15 cm

A_2 = cross-sectional area of throat = $\frac{\pi}{4}d_2^2 = \frac{\pi}{4}(15)^2 = 176.625 \text{ cm}^2$

P_1 = pressure at inlet = 15 $N/cm^2 = 15 \times 10^4$ N/m^2

ρ = density of water = 1000 kg/m^3

$\therefore \quad \frac{P_1}{\rho g} = \frac{15 \times 10^4}{10^3 \times 9.81}$; 15 m of water

$\frac{P_2}{\rho g} = -40$ cm of mercury

$\Rightarrow \frac{P_2}{\rho g} = -0.40$ m of mercury

$\Rightarrow \frac{P_2}{\rho g} = -0.40 \times 13.6 = -5.44$ m of water

$\therefore$ Difference of pressure heads $= h = \dfrac{P_1}{\rho g} - \dfrac{P_2}{\rho g}$

$$\Rightarrow h = 15 + 5.44 = 20.44 \text{ m of water}$$
$$\Rightarrow h = 2044 \text{ cm of water}$$

The discharge through ventrurimeter is

$$Q = C_d \frac{A_1 A_2}{\sqrt{A_1^2 - A_2^2}} \sqrt{2gh}$$

$$\Rightarrow Q = 0.98 \times \frac{706.5 \times 176.625}{\sqrt{(706.5)^2 - (176.625)^2}} \sqrt{2 \times 9.81 \times 2044}$$

$$\Rightarrow Q = 0.98 \times \frac{124785.5625}{684.1} \sqrt{40103}$$
$$\Rightarrow Q = 0.98 \times 182.4 \times 200.26$$
$$\Rightarrow Q = 35797 \text{ cm}^3 / \text{sec}$$
$$\Rightarrow Q = 35.797 \text{ Lt/sec}$$

EXAMPLE 9. A 30 cm×15 cm ventrurimeter is inserted in a vertical pipe carrying water, flowing in upward direction. A differential mercury manometer connected to the inlet and throat gives a reading of 30 cm. Determine the discharge provided the co-efficient of discharge (C_d) is 0.98.

Ans: In this case,

Let d_1 = diameter of inlet = 30 cm

A_1 = cross-sectional area of inlet $= \dfrac{\pi}{4} d_1^2 = \dfrac{\pi}{4}(30)^2 = 706.5 \text{ cm}^2$

d_2 = diameter of throat = 15 cm

A_2 = cross-sectional area of throat $= \dfrac{\pi}{4} d_2^2 = \dfrac{\pi}{4}(15)^2 = 176.625 \text{ cm}^2$

C_d = co-efficient of discharge = 0.98

ρ_m = density of mercury in manometer = 13545.848 kg/m^3

ρ_w = density of water flows through pipe = 1000 kg/m^3

y = difference of the mercury level in manometric U-tube = 30 cm

The difference of pressure head is given by

$$h = y\left[\frac{\rho_m}{\rho_w} - 1\right] = 30\left[\frac{13545.848}{1000} - 1\right]$$
$$\Rightarrow h = 30(12.545)$$
$$\Rightarrow h = 376.35 \text{ cm of water}$$

Discharge through ventrurimeter is

$$Q = C_d \frac{A_1 A_2}{\sqrt{A_1^2 - A_2^2}}\sqrt{2gh}$$

$$\Rightarrow Q = 0.98 \times \frac{706.5 \times 176.625}{\sqrt{(706.5)^2 - (176.625)^2}}\sqrt{2 \times 9.81 \times 376.35}$$

$$\Rightarrow Q = 0.98 \times 182.4 \times \sqrt{7384}$$

$$\Rightarrow Q = 0.98 \times 182.4 \times 86 = 15372.672 \text{ cm}^3/\text{sec}$$

$$\Rightarrow Q = 15.372 \text{ Lt/sec}$$

EXAMPLE 10. A 30 cm×15 cm ventrurimeter is inserted in a vertical pipe carrying oil of density 800 kg/m^3, flowing in upward direction. A differential mercury manometer connected to the inlet and throat gives a reading of 30 cm. Determine the discharge provided the co-efficient of discharge (C_d) is 0.98.

Ans: In this case,
Let d_1 = diameter of inlet = 30 cm

A_1 = cross-sectional area of inlet = $\dfrac{\pi}{4}d_1^2 = \dfrac{\pi}{4}(30)^2 = 706.5$ cm^2

d_2 = diameter of throat = 15 cm

A_2 = cross-sectional area of throat = $\dfrac{\pi}{4}d_2^2 = \dfrac{\pi}{4}(15)^2 = 176.625$ cm^2

C_d = co-efficient of discharge = 0.98

ρ_m = density of mercury in manometer = 13545.848 kg/m^3

ρ_o = density of oil flows through pipe = 800 kg/m^3

y = difference of the mercury level in manometric U-tube = 30 cm

The difference of pressure head is given by

$$h = y\left[\frac{\rho_m}{\rho_o}-1\right] = 30\left[\frac{13545.848}{800}-1\right]$$

$$\Rightarrow h = 30(15.93)$$
$$\Rightarrow h = 478 \text{ cm of water}$$

Discharge through ventrurimeter is

$$Q = C_d \frac{A_1 A_2}{\sqrt{A_1^2 - A_2^2}}\sqrt{2gh}$$

$$\Rightarrow Q = 0.98 \times \frac{706.5 \times 176.625}{\sqrt{(706.5)^2 - (176.625)^2}}\sqrt{2 \times 9.81 \times 478}$$

$$\Rightarrow Q = 0.98 \times 182.4 \times \sqrt{9378}$$

$$\Rightarrow Q = 0.98 \times 182.4 \times 96.84 = 17310.34 \text{ cm}^3/\text{sec}$$

$$\Rightarrow Q = 17.31 \text{ Lt/sec}$$

EXAMPLE 11. The water is flowing through a pipe of diameter 30 cm. The pipe is inclined and a ventrurimeter is inserted in the pipe. The diameter of ventrurimeter at throat is 15 cm as shown in figure. The difference of pressure between the inlet and the throat of the ventrurimeter is measured by a fluid of density 800 kg / m^3 in an inverted U-tube which gives a reading of 40 cm. the loss of head between the inlet and throat is 0.3 times the kinetic head of the pipe. Determine the discharge provided the co-efficient of discharge (C_d) is 0.98.

Ans: In this case,

Let d_1 = diameter of inlet = 30 cm

A_1 = cross-sectional area of inlet = $\dfrac{\pi}{4} d_1^2 = \dfrac{\pi}{4}(30)^2 = 706.5$ cm^2

d_2 = diameter of throat = 15 cm

A_2 = cross-sectional area of throat = $\dfrac{\pi}{4} d_2^2 = \dfrac{\pi}{4}(15)^2 = 176.625$ cm^2

C_d = co-efficient of discharge = 0.98

ρ_m = density of mercury in manometer = 13545.848 kg / m^3

ρ_f = density of fluid flows through pipe = 800 kg / m^3

y = difference of the mercury level in manometric U-tube = 40 cm

The difference of pressure head is given by

$$h = y\left[\frac{\rho_m}{\rho_f} - 1\right] = 40\left[\frac{13545.848}{800} - 1\right]$$

$$\Rightarrow h = 40(15.93)$$

$$\Rightarrow h = 637.2 \text{ cm of water}$$

$$\therefore \quad \left(\frac{P_1}{\rho g} + z_1\right) - \left(\frac{P_2}{\rho g} + z_2\right) = h = 637.2$$

Loss of head, $h_L = 0.3 \times$ kinetic head of pipe $= 0.3 \times \dfrac{v_1^2}{2g}$

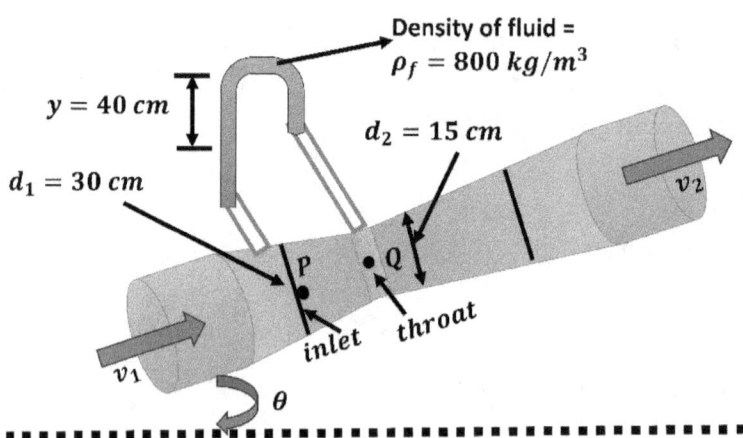

Now applying Bernoulli's equation along the streamline from the point P to point Q, we obtain as

$$\frac{P_1}{\rho g}+\frac{v_1^2}{2g}+z_1 = \frac{P_2}{\rho g}+\frac{v_2^2}{2g}+z_2+h_L$$

$$\Rightarrow \left(\frac{P_1}{\rho g}+z_1\right)-\left(\frac{P_2}{\rho g}+z_2\right)+\frac{v_1^2}{2g}-\frac{v_2^2}{2g}=h_L$$

$$\Rightarrow h+\frac{v_1^2}{2g}-\frac{v_2^2}{2g}=h_L$$

$$\Rightarrow 637.2+\frac{v_1^2}{2g}-\frac{v_2^2}{2g}=0.3\times\frac{v_1^2}{2g}$$

$$\Rightarrow 637.2+0.7\times\frac{v_1^2}{2g}-\frac{v_2^2}{2g}=0 \qquad \text{-------(1)}$$

Now apply continuity equation along the streamline from the point P to point Q, we obtain as

$$A_1 v_1 = A_2 v_2$$

$$\Rightarrow v_1 = \frac{A_2}{A_1} v_2$$

$$\Rightarrow v_1 = \frac{176.625}{706.5} v_2$$

$$\Rightarrow v_1 = \frac{v_2}{4}$$

Using above relation, equation (1) reduced to

$$637.2 + \frac{0.7}{2g}\left(\frac{v_2}{4}\right)^2 - \frac{v_2^2}{2g} = 0$$

$$\Rightarrow 637.2 + \frac{v_2^2}{2g}\left[\frac{0.7}{16} - 1\right] = 0$$

$$\Rightarrow \frac{v_2^2}{2g}(-0.96) = -637.2$$

$$\Rightarrow v_2^2 = \frac{637.2 \times 2 \times 9.81}{0.96} ; \ 13022.775$$

$$\Rightarrow v_2 = \sqrt{13022.775} ; \ 114 \ \text{cm/sec}$$

Discharge through ventrurimeter is

$$Q = A_2 v_2$$

$$\Rightarrow Q = 176.625 \times 114$$
$$\Rightarrow Q = 20135.25 \ \text{cm}^3 / \text{sec}$$
$$\Rightarrow Q = 20.13 \ \text{Lt/sec}$$

EXAMPLE 12. A 30 cm×15 cm ventrurimeter is provided in a vertical pipe line carrying oil of density 800 kg/m^3, the flow in upward direction. The difference in elevation of the throat section and the inlet section of the ventrurimeter is 50 cm as shown in figure. The differential U-tube mercury manometer shows a gauge deflection of 40 cm. Determine the pressure difference between the inlet and throat provided the co-efficient of discharge (C_d) is 0.98.

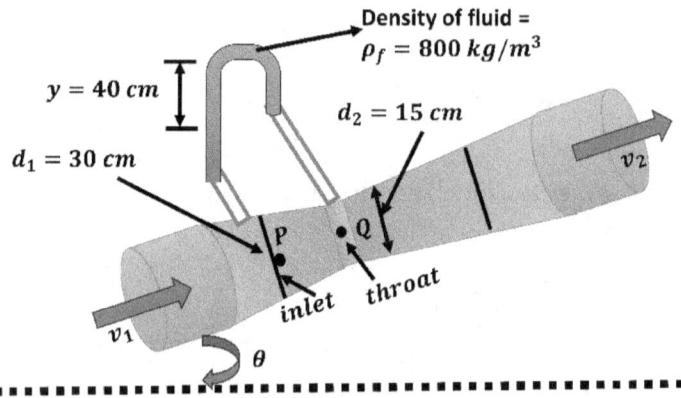

Density of fluid = $\rho_f = 800\ kg/m^3$

$y = 40\ cm$

$d_2 = 15\ cm$

$d_1 = 30\ cm$

v_2

P Q

v_1

inlet   throat

θ

Ans: In this case,

Let d_1 = diameter of inlet = 30 cm

A_1 = cross-sectional area of inlet = $\dfrac{\pi}{4}d_1^2 = \dfrac{\pi}{4}(30)^2 = 706.5\ \text{cm}^2$

d_2 = diameter of throat = 15 cm

A_2 = cross-sectional area of throat = $\dfrac{\pi}{4}d_2^2 = \dfrac{\pi}{4}(15)^2 = 176.625\ \text{cm}^2$

C_d = co-efficient of discharge = 0.98

ρ_m = density of mercury in manometer = 13545.848 kg/m^3

ρ_o = density of oil flows through pipe = 800 kg/m^3

y = difference of the mercury level in manometric U-tube = 40 cm

The difference of pressure head is given by

$$h = y\left[\frac{\rho_m}{\rho_o}-1\right] = 40\left[\frac{13545.848}{800}-1\right]$$
$$\Rightarrow h = 40(15.93)$$
$$\Rightarrow h = 637.2\ \text{cm of water}$$

$$\therefore \quad \left(\frac{P_1}{\rho g} + z_1\right) - \left(\frac{P_2}{\rho g} + z_2\right) = h = 637.2$$

$$\Rightarrow \left(\frac{P_1}{\rho g} - \frac{P_2}{\rho g}\right) + z_1 - z_2 = 637.2$$

$$\Rightarrow \left(\frac{P_1}{\rho g} - \frac{P_2}{\rho g}\right) + 50 = 637.2$$

$$\Rightarrow \left(\frac{P_1}{\rho g} - \frac{P_2}{\rho g}\right) = 587.2$$

$$\Rightarrow P_1 - P_2 = 587.2 \times \rho g$$

$$\Rightarrow P_1 - P_2 = 587.2 \times 800 \times 9.81$$

$$\Rightarrow P_1 - P_2 = 4608346 \text{ N/m}^2$$

$$\Rightarrow P_1 - P_2 = \frac{4608346}{10^4} = 460.83 \text{ N/cm}^2$$

EXAMPLE 13. In a 200 mm diameter horizontal pipe, a ventrurimeter of 0.5 contraction ratio sustain. The head of water on the ventrurimeter when there is no flow is 3 m (gauge). The pressure at throat is 4 m of water and atmospheric pressure head is 10.3 m of water. Determine the discharge of flow provided the co-efficient of discharge (C_d) is 0.97.

Ans: In this case,

d_1 = diameter of pipeline = 200 mm = 20 cm

A_1 = cross-sectional area of inlet $= \dfrac{\pi}{4}d_1^2 = \dfrac{\pi}{4}(20)^2 = 314 \text{ cm}^2$

d_2 = diameter of throat = $0.5 \times d_1 = 0.5 \times 20 = 10$ cm

A_2 = cross-sectional area of throat $= \dfrac{\pi}{4}d_2^2 = \dfrac{\pi}{4}(10)^2 = 78.5 \; cm^2$

Pressure at inlet of pipeline $= \dfrac{P_1}{\rho g} = 3 \text{ m} = 3 + 10.3 = 13.3(absoulte)$

Pressure at throat $= \dfrac{P_2}{\rho g} = 4$

$\therefore$ Difference of pressure head, $h = \dfrac{P_1}{\rho g} - \dfrac{P_2}{\rho g}$

$$\Rightarrow h = \frac{P_1}{\rho g} - \frac{P_2}{\rho g} = 13.3 - 4 = 9.3 \text{ m} = 930 \text{ m}$$

Discharge through ventrurimeter is

$$Q = C_d \frac{A_1 A_2}{\sqrt{A_1^2 - A_2^2}} \sqrt{2gh}$$

$$\Rightarrow Q = 0.97 \times \frac{314 \times 78.5}{\sqrt{(314)^2 - (78.5)^2}} \times \sqrt{2 \times 9.81 \times 930}$$

$$\Rightarrow Q = 0.97 \times \frac{24649}{\sqrt{98596 - 6162.25}} \times \sqrt{18246.6}$$

$$\Rightarrow Q = 0.97 \times \frac{24649}{\sqrt{92433.75}} \times 13.5$$

$$\Rightarrow Q = 0.97 \times \frac{24649}{304} \times 13.5$$

$$\Rightarrow Q = 0.97 \times 81.08 \times 13.5$$

$$\Rightarrow Q = 1062 \text{ cm}^3 / \sec = 1.06 \text{ Lt/sec}$$

EXAMPLE 14. Determine the discharge of flow through a ventrurimeter fitted in a pipeline of diameter 30 cm. The ratio of diameter of throat and inlet of the ventrurimeter is 1:3. The pressure at the inlet of the ventrurimeter is 14 N/cm^2 and vacuum pressure at the throat is 40 cm of the mercury provided the co-efficient of discharge (C_d) is 0.98.

Ans: In this case,

d_1 = diameter of inlet = 30 cm

A_1 = cross-sectional area of inlet $= \dfrac{\pi}{4}d_1^2 = \dfrac{\pi}{4}(30)^2 = 706.5$ cm^2

d_2 = diameter of throat = 10 cm

A_2 = cross-sectional area of throat $= \dfrac{\pi}{4}d_2^2 = \dfrac{\pi}{4}(10)^2 = 78.5$ cm^2

P_1 = pressure at inlet = 14 $N/cm^2 = 14 \times 10^4$ N/m^2

ρ = density of water $= 1000$ kg/m^3

$\therefore \quad \dfrac{P_1}{\rho g} = \dfrac{14 \times 10^4}{10^3 \times 9.81}$; 14 m of water

$\dfrac{P_2}{\rho g} = -40$ cm of mercury

$\Rightarrow \dfrac{P_2}{\rho g} = -0.40$ m of mercury

$\Rightarrow \dfrac{P_2}{\rho g} = -0.40 \times 13.6 = -5.44$ m of water

$\therefore \qquad$ Difference of pressure heads $= h = \dfrac{P_1}{\rho g} - \dfrac{P_2}{\rho g}$

$\Rightarrow h = 18 + 5.44 = 23.44$ m of water

$\Rightarrow h = 2344$ cm of water

The discharge through ventrurimeter is

$$Q = C_d \dfrac{A_1 A_2}{\sqrt{A_1^2 - A_2^2}}\sqrt{2gh}$$

$$\Rightarrow Q = 0.98 \times \dfrac{706.5 \times 78.5}{\sqrt{(706.5)^2 - (78.5)^2}}\sqrt{2 \times 9.81 \times 2344}$$

$$\Rightarrow Q = 0.98 \times \frac{55460.25}{702.126} \sqrt{43320.96}$$

$$\Rightarrow Q = 0.98 \times 78.989 \times 208.14$$

$$\Rightarrow Q = 16111 \text{ cm}^3 / \text{sec}$$

$$\Rightarrow Q = 16.11 \text{ Lt/sec}$$

EXAMPLE 15. A ventrurimeter is used to measure the discharge of water in horizontal pipe line of diameter 30 cm. If the ratio of diameter of throat and inlet is 1:2, the difference in pressure between the throat and inlet is equal to 3 m head of water and loss of head through gauge is $\dfrac{1}{8}$ of the throat velocity head. Determine the discharge of flow provided the co-efficient of discharge (C_d) is 0.98.

Ans: In this case,

Let d_1 = diameter of inlet = 30 cm

A_1 = cross-sectional area of inlet = $\dfrac{\pi}{4} d_1^2 = \dfrac{\pi}{4}(30)^2 = 706.5 \text{ cm}^2$

d_2 = diameter of throat = 15 cm

A_2 = cross-sectional area of throat = $\dfrac{\pi}{4} d_2^2 = \dfrac{\pi}{4}(15)^2 = 176.625 \text{ cm}^2$

C_d = co-efficient of discharge = 0.98

h = difference of pressure head = 3 m = 300 cm

The discharge through ventrurimeter is

$$Q = C_d \frac{A_1 A_2}{\sqrt{A_1^2 - A_2^2}} \sqrt{2gh}$$

$$\Rightarrow Q = 0.98 \times \frac{706.5 \times 176.625}{\sqrt{\left(706.5\right)^2 - \left(176.625\right)^2}} \times \sqrt{2 \times 9.81 \times 300}$$

$$\Rightarrow Q = 0.98 \times 182.4 \times 76.72$$

$$\Rightarrow Q = 13714 \text{ cm}^3 / \sec$$

$$\Rightarrow Q = 13.714 \text{ Lt/sec}$$

EXAMPLE 16. An oil of density 800 kg/m^3 is flowing upwards at the rate of 0.08 $m^3/\sec$, through a vertical ventrurimeter with an inlet diameter of 300 mm and throat diameter of 150 mm. The vertical distance between pressure tappings is 300 mm. Calculate the difference in level of the mercury columns of the differential manometer which is connected to the tappings, in place of pressure gauges provided the co-efficient of discharge (C_d) is 0.98.

Ans: In this case,

d_1 = diameter of inlet = 300 mm = 0.3 m

A_1 = cross-sectional area of inlet = $\dfrac{\pi}{4} d_1^2 = \dfrac{\pi}{4}(0.3)^2 = 0.07065 \text{ m}^2$

d_2 = diameter of throat section = 150 mm = 0.15 m

A_2 = cross-sectional area of throat = $\dfrac{\pi}{4} d_2^2 = \dfrac{\pi}{4}(0.15)^2 = 0.0176625 \text{ m}^2$

Q = discharge of oil = 0.08 $m^3/\sec$

C_d = co-efficient of discharge = 0.98

ρ_m = density of mercury = 13545.848 kg/m^3

ρ_o = density of oil = 800 kg/m^3

h = difference of pressure heads = 300 mm

Let y be difference of the mercury level in manometric U-tube

$$\Rightarrow y \left[\frac{\rho_m}{\rho_o} - 1 \right] = 300$$

$$\Rightarrow y \left[\frac{13545.848}{800} - 1 \right] = 300$$

$$\Rightarrow y (15.93231) = 300$$

$$\Rightarrow y = \frac{300}{15.93231} = 18.83 \text{ mm} = 0.01883 \text{ m}$$

EXAMPLE 17. A ventrurimeter is installed in a 300 mm diameter horizontal pipe line used to measure the flow of water. The throat pipe rate is $\frac{1}{4}$. The pressure in the pipe line is 14 N/cm^2 and vacuum pressure at the throat is 36 cm of mercury. Determine the discharge rate of the flow in the pipe line provided the co-efficient of discharge (C_d) is 0.98.

Ans: In this case,

d_1 = diameter of inlet $= 300$ mm $= 30$ cm

A_1 = cross-sectional area of inlet $= \frac{\pi}{4} d_1^2 = \frac{\pi}{4} (30)^2 = 706.5 \text{ cm}^2$

d_2 = diameter of throat $= \frac{1}{4} \times d_1 = \frac{1}{4} \times 30 = 7.5$ cm

A_2 = cross-sectional area of throat $= \frac{\pi}{4} d_2^2 = \frac{\pi}{4} (7.5)^2 = 44.156 \text{ cm}^2$

P_1 = pressure at inlet $= 14 \ N/cm^2 = 14 \times 10^4$ N/m^2

ρ = density of water $= 1000$ kg/m^3

$\therefore \quad \frac{P_1}{\rho g} = \frac{14 \times 10^4}{10^3 \times 9.81}$; 14 m of water

$$\frac{P_2}{\rho g} = -36 \text{ cm of mercury}$$

$$\Rightarrow \frac{P_2}{\rho g} = -0.36 \text{ m of mercury}$$

$$\Rightarrow \frac{P_2}{\rho g} = -0.36 \times 13.6 = -4.896 \text{ m of water}$$

$\therefore$ Difference of pressure heads $= h = \dfrac{P_1}{\rho g} - \dfrac{P_2}{\rho g}$

$$\Rightarrow h = 18 + 4.896 = 22.896 \text{ m of water}$$
$$\Rightarrow h = 2289.6 \text{ cm of water}$$

The discharge through ventrurimeter is

$$Q = C_d \frac{A_1 A_2}{\sqrt{A_1^2 - A_2^2}} \sqrt{2gh}$$

$$\Rightarrow Q = 0.98 \times \frac{706.5 \times 44.156}{\sqrt{(706.5)^2 - (44.156)^2}} \sqrt{2 \times 9.81 \times 2289.6}$$

$$\Rightarrow Q = 0.98 \times \frac{31196.214}{\sqrt{497192.49766}} \times \sqrt{44921.952}$$

$$\Rightarrow Q = 0.98 \times \frac{31196.214}{705.118} \times 212$$

$$\Rightarrow Q = 0.98 \times 44.24 \times 212$$

$$\Rightarrow Q = 9191.3 \text{ cm}^3 / \text{sec}$$

$$\Rightarrow Q = 9.1913 \text{ Lt/sec}$$

EXAMPLE 18. The maximum flow through a 300 mm diameter horizontal pipe line is 18200 $Lt/\min$. A ventrurimeter is installed at a point of the pipe line where the pressure head is 4.6 m of water. Calculate the smallest diameter of throat so that pressure at the throat is never negative. Assume that co-efficient of discharge is 1.

Ans: In this case,

d_1 = diameter of inlet = 300 mm = 0.3 m

A_1 = cross-sectional area of inlet = $\dfrac{\pi}{4}d_1^2 = \dfrac{\pi}{4}(0.3)^2 = 0.7065$ m^2

Let d_2 = smallest diameter of throat

C_d = co-efficient of discharge = 1

Q = discharge rate of flow

= 18200 Lt/min = $\dfrac{18200}{60}$ Lt/sec = 303.33 Lt/sec = 0.3033 m^3 / sec

$\dfrac{P_1}{\rho g}$ = pressure head at inlet = 4.6 m of water

$\dfrac{P_2}{\rho g}$ = pressure head at throat = 0

$\therefore$ Difference of pressure heads = $h = \dfrac{P_1}{\rho g} - \dfrac{P_2}{\rho g} = 4.6$ m

The discharge through ventrurimeter is

$$Q = C_d \frac{A_1 A_2}{\sqrt{A_1^2 - A_2^2}} \sqrt{2gh}$$

$$\Rightarrow 0.3033 = 1 \times \frac{A_1 A_2}{\sqrt{A_1^2 - A_2^2}} \sqrt{2 \times 9.81 \times 4.6}$$

$$\Rightarrow \frac{A_1 A_2}{\sqrt{A_1^2 - A_2^2}} = \frac{0.3033}{\sqrt{2 \times 9.81 \times 4.6}} = \frac{0.3033}{9.5}$$

$$\Rightarrow \frac{A_1 A_2}{\sqrt{A_1^2 - A_2^2}} ;\ 0.032$$

$$\Rightarrow A_1^2 A_2^2 = (0.32)^2 \left(A_1^2 - A_2^2 \right)$$

$$\Rightarrow A_2^2 = \frac{(0.032)^2}{(0.7065)^2} \left[(0.7065)^2 - A_2^2 \right]$$

$$\Rightarrow A_2^2 = \frac{0.001024}{0.49914225} \left[(0.7065)^2 - A_2^2 \right]$$

$$\Rightarrow A_2^2 = 0.0021 \left[(0.7065)^2 - A_2^2 \right]$$

$$\Rightarrow A_2^2 = 0.001024 - 0.0021 A_2^2$$

$$\Rightarrow A_2^2 (1 + 0.0021) = 0.001024$$

$$\Rightarrow A_2^2 = \frac{0.001024}{1.0021} = 0.001022$$

$$\Rightarrow A_2 = \sqrt{0.001022} \; ; \; 0.032$$

$$\Rightarrow \frac{\pi}{4} d_2^2 = 0.032$$

$$\Rightarrow d_2^2 = \frac{0.032 \times 4}{3.14} = 0.0408$$

$$\Rightarrow d_2 = \sqrt{0.0408} = 0.202 \text{ m} = 202 \text{ mm}$$

EXAMPLE 19. A ventrurimeter of inlet diameter 300 mm and throat diameter 150 mm is fixed in a vertical pipe line. A fluid of density 800 kg/m^3 is flowing through pipe line in upward direction. A differential manometer gives reading of 100 mm when installed between inlet and throat. The vertical difference between inlet and throat is 500 mm provided the co-efficient of discharge (C_d) is 0.98. determine the pressure difference between inlet and throat.

Ans: In this case,

Let d_1 = diameter of inlet = 300 mm = 30 cm

A_1 = cross-sectional area of inlet = $\frac{\pi}{4} d_1^2 = \frac{\pi}{4}(30)^2 = 706.5 \text{ cm}^2$

d_2 = diameter of throat = 150 mm = 15 cm

A_2 = cross-sectional area of throat = $\dfrac{\pi}{4} d_2^2 = \dfrac{\pi}{4}(15)^2 = 176.625$ cm^2

C_d = co-efficient of discharge = 0.98

ρ_m = density of mercury in manometer = 13545.848 kg / m^3

ρ_f = density of fluid flows through pipe = 800 kg / m^3

y = difference of the mercury level in manometric U-tube = 100 mm = 10 cm

The difference of pressure head is given by

$$h = y\left[\frac{\rho_m}{\rho_f} - 1\right] = 10\left[\frac{13545.848}{800} - 1\right]$$

$$\Rightarrow h = 10(15.83)$$

$$\Rightarrow h = 150.83 \text{ cm of water}$$

$$\therefore \quad \left(\frac{P_1}{\rho g} + z_1\right) - \left(\frac{P_2}{\rho g} + z_2\right) = h = 150.83$$

$$\Rightarrow \left(\frac{P_1}{\rho g} - \frac{P_2}{\rho g}\right) + z_1 - z_2 = 150.83$$

$$\Rightarrow \left(\frac{P_1}{\rho g} - \frac{P_2}{\rho g}\right) + 50 = 150.83$$

$$\Rightarrow \left(\frac{P_1}{\rho g} - \frac{P_2}{\rho g}\right) = 100.83$$

$$\Rightarrow P_1 - P_2 = 100.83 \times \rho g$$

$$\Rightarrow P_1 - P_2 = 100.83 \times 800 \times 9.81$$

$$\Rightarrow P_1 - P_2 = 791313.84 \text{ N/m}^2$$

$$\Rightarrow P_1 - P_2 = \frac{791313.84}{10^4} = 79.131 \text{ N/cm}^2$$

EXAMPLE 20. A ventrurimeter is used to measure the discharge of water in horizontal pipe line of diameter 30 cm. If the ratio of diameter of throat and inlet is 1:2, The pressure at the inlet is 70 kPa and it is desired that the pressure at any point should not fall below 2.5 m of absolute water. Determine the maximum discharge rate of water through ventrurimeter provided co-efficient of discharge (C_d) is 0.98 and atmospheric pressure is taken as 100 kPa.

Ans: In this case,

d_1 = diameter of pipeline $= 30$ cm

A_1 = cross-sectional area of inlet $= \dfrac{\pi}{4}d_1^2 = \dfrac{\pi}{4}(30)^2 = 706.5$ cm^2

d_2 = diameter of throat $= 15$ cm

A_2 = cross-sectional area of throat $= \dfrac{\pi}{4}d_2^2 = \dfrac{\pi}{4}(15)^2 = 176.625$ cm^2

C_d = co-efficient of discharge $= 0.98$

Pressure at inlet $= P_1 = 70$ kPa $= 70 \times 10^3$ N/m^2 (gauge)

P_{atm} = atmospheric pressure $= 100$ kPa $= 100 \times 10^3$ N/m^2

$\therefore \quad P_1 = 70 \times 10^3 + 100 \times 10^3$ N/m$^2 = 170 \times 10^3$ (absoulte)

Pressure head at inlet of pipeline

$= \dfrac{P_1}{\rho g} = \dfrac{170 \times 10^3}{1000 \times 9.81} = 17.33$ m of water

Pressure at throat $= 2.5$ m $(absolute)$

$\therefore \qquad$ Difference of pressure head, $h = \dfrac{P_1}{\rho g} - \dfrac{P_2}{\rho g}$

$$\Rightarrow h = \dfrac{P_1}{\rho g} - \dfrac{P_2}{\rho g} = 17.33 - 2.5$$
$$\Rightarrow h = 14.83 \text{ m of water}$$
$$\Rightarrow h = 1483 \text{ cm}$$

Discharge through ventrurimeter is

$$Q = C_d \frac{A_1 A_2}{\sqrt{A_1^2 - A_2^2}} \sqrt{2gh}$$

$$\Rightarrow Q = 0.97 \times \frac{706.5 \times 176.625}{\sqrt{(706.5)^2 - (176.625)^2}} \times \sqrt{2 \times 9.81 \times 1483}$$

$$\Rightarrow Q = 0.98 \times 182.4 \times \sqrt{29096.46}$$
$$\Rightarrow Q = 0.98 \times 182.4 \times 170.58$$
$$\Rightarrow Q = 30492 \text{ cm}^3 / \sec = 30.482 \text{ Lt/sec}$$

EXAMPLE 21. Determine the discharge rate of water flow through a 30 cm diameter pipe line installed in inclined position where ventrurimeter is inserted whose throat diameter is of 15 cm as shown in figure. The difference of pressure between the inlet and throat measured by an inverted U-tube filled with a fluid of density 400 kg/m^3 which gives the reading of 30 cm. the loss of head between the inlet and throat is 0.3 times the kinetic head of the pipe line.

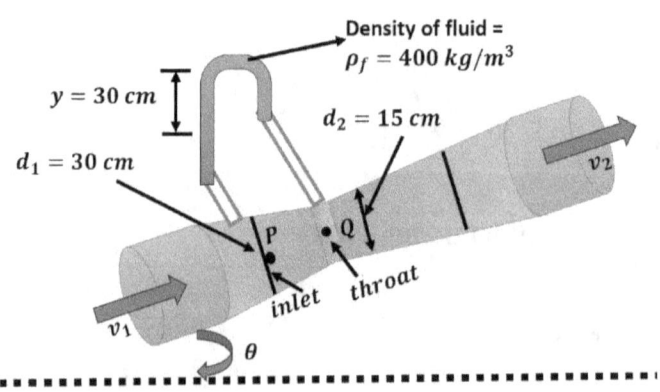

Ans: In this case,

Let d_1 = diameter of inlet = 30 cm

A_1 = cross-sectional area of inlet = $\frac{\pi}{4} d_1^2 = \frac{\pi}{4}(30)^2 = 706.5 \text{ cm}^2$

d_2 = diameter of throat = 15 cm

A_2 = cross-sectional area of throat $= \dfrac{\pi}{4}d_2^2 = \dfrac{\pi}{4}(15)^2 = 176.625$ cm^2

C_d = co-efficient of discharge $= 0.98$

ρ_m = density of mercury in manometer $= 13545.848$ kg/m^3

ρ_w = density of water flows through pipe $= 400$ kg/m^3

y = difference of the mercury level in manometric U-tube $= 30$ cm

The difference of pressure head is given by

$$h = y\left[\frac{\rho_m}{\rho_o} - 1\right] = 30\left[\frac{13545.848}{400} - 1\right]$$
$$\Rightarrow h = 30(33.86)$$
$$\Rightarrow h = 1015.8 \text{ cm of water}$$

$$\therefore \quad \left(\frac{P_1}{\rho g} + z_1\right) - \left(\frac{P_2}{\rho g} + z_2\right) = h = 1015.8$$

Loss of head, $h_L = 0.3 \times$ kinetic head of pipe $= 0.3 \times \dfrac{v_1^2}{2g}$

Now applying Bernoulli's equation along the streamline from the point P to point Q, we obtain as

$$\frac{P_1}{\rho g} + \frac{v_1^2}{2g} + z_1 = \frac{P_2}{\rho g} + \frac{v_2^2}{2g} + z_2 + h_L$$

$$\Rightarrow \left(\frac{P_1}{\rho g} + z_1\right) - \left(\frac{P_2}{\rho g} + z_2\right) + \frac{v_1^2}{2g} - \frac{v_2^2}{2g} = h_L$$

$$\Rightarrow h + \frac{v_1^2}{2g} - \frac{v_2^2}{2g} = h_L$$

$$\Rightarrow 1015.8 + \frac{v_1^2}{2g} - \frac{v_2^2}{2g} = 0.3 \times \frac{v_1^2}{2g}$$

$$\Rightarrow 1015.8 + 0.7 \times \frac{v_1^2}{2g} - \frac{v_2^2}{2g} = 0 \qquad \text{-------(1)}$$

Now apply continuity equation along the streamline from the point P to point Q, we obtain as

$$A_1 v_1 = A_2 v_2$$

$$\Rightarrow v_1 = \frac{A_2}{A_1} v_2$$

$$\Rightarrow v_1 = \frac{176.625}{706.5} v_2$$

$$\Rightarrow v_1 = \frac{v_2}{4}$$

Using above relation, equation (1) reduced to

$$1015.8 + \frac{0.7}{2g}\left(\frac{v_2}{4}\right)^2 - \frac{v_2^2}{2g} = 0$$

$$\Rightarrow 1015.8 + \frac{v_2^2}{2g}\left[\frac{0.7}{16} - 1\right] = 0$$

$$\Rightarrow \frac{v_2^2}{2g}(-0.96) = -1015.8$$

$$\Rightarrow v_2^2 = \frac{1015.8 \times 2 \times 9.81}{0.96} ;\ 20760$$

$$\Rightarrow v_2 = \sqrt{20760} ;\ 144.08 \text{ cm/sec}$$

Discharge through ventrurimeter is

$$Q = A_2 v_2$$

$$\Rightarrow Q = 176.625 \times 144.08$$

$$\Rightarrow Q = 25448.13 \text{ cm}^3 / \text{sec}$$

$$\Rightarrow Q = 25.44 \text{ Lt/sec}$$

ORIFICEMETER

CONSTRUCTION: Orificemeter comprise of four parts namely, inlet suction, orifice plate, flow conditioner, and outlet section. The inlet section through which fluid inflow into orificemeter and outflow through outlet section. The thin sized plate called orifice plate contain few holes through which fluid will pass and there is a pressure drop in fluid flow. The system attached a flow conditioner which increase the flow through inlet section.

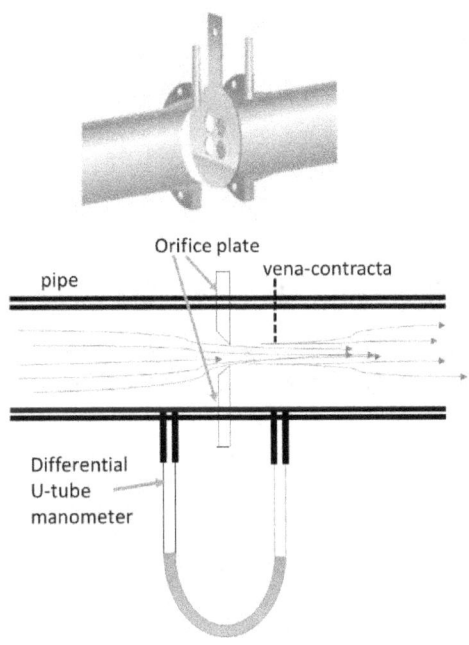

WORKING: The orifice plate is used for give a resistance to the fluid flow which results in the pressure drop in the flowing fluid. The working principle of orificemeter is phenomenon of Bernoulli's theorem. The drop in pressure can be measured by a differential manometer.

The length of Orifice would be within the range of 10-800 mm and diameter of Orifice plate may be within the range of 0.4 to 0.8 times the diameter of the pipe. The pressure of fluid within the system can be maximum up to 400 Bar.

Consider a Orificemeter positioned at an inclined line to the horizon to measure the flow rate through the pipe as shown in figure. Let us consider a steady, incompressible fluid such that the velocity and pressure at any section will be uniform. Then, the velocity and pressure at the converging part (point 1) and at vena-contracta (point 2) are v_1, P_1 and v_2, P_2. Here the vertical column heights at the point 1 and point 2 be h_1

and h_2 ($h_2 > h_1$) so that assume that $\Delta h = h_2 - h_1$.
Next, applying the Bernoulli's equation along the streamline from the point 1 to the point 2, we obtain as:

$$\frac{P_1}{\rho g} + \frac{v_1^2}{2g} + h_1 = \frac{P_2}{\rho g} + \frac{v_2^2}{2g} + h_2$$

$$\Rightarrow \frac{1}{2}(v_2^2 - v_1^2) = \frac{1}{\rho}(P_1 - P_2) + g(h_1 - h_2)$$

$$\Rightarrow \frac{1}{2}(v_2^2 - v_1^2) = \frac{1}{\rho}(P_1 - P_2) - g(h_2 - h_1) \qquad \text{-----(1)}$$

According to the continuity principle, we have

$$A_1 v_1 = A_2 v_2 \Rightarrow v_1 = \frac{A_2}{A_1} v_2 \quad \text{Now, the equ (1) becomes}$$

$$\frac{1}{2}\left[v_2^2 - v_2^2\left(\frac{A_2}{A_1}\right)^2 \right] = \frac{1}{\rho}(P_1 - P_2) - g(h_2 - h_1)$$

$$\Rightarrow \frac{v_2^2}{2}\left[1 - \left(\frac{A_2}{A_1}\right)^2 \right] = \frac{1}{\rho}(P_1 - P_2) - g(h_2 - h_1)$$

$$\Rightarrow v_2 = \sqrt{\frac{2\left[\dfrac{(P_1 - P_2)}{\rho} - g(h_2 - h_1) \right]}{\left[1 - \left(\dfrac{A_2}{A_1}\right)^2 \right]}}$$

Thus, Q_2 be the volumetric discharge of the fluid through the pipe which is defined as

$$Q_2 = v_2 A_2$$

$$\Rightarrow Q_2 = A_2 \sqrt{\dfrac{2\left[\dfrac{(P_1-P_2)}{\rho}-g(h_2-h_1)\right]}{\left[1-\left(\dfrac{A_2}{A_1}\right)^2\right]}}$$

$$\Rightarrow Q_2 = \dfrac{A_2}{\sqrt{\left[1-\left(\dfrac{A_2}{A_1}\right)^2\right]}}\sqrt{2\left[\dfrac{(P_1-P_2)}{\rho}-g(h_2-h_1)\right]} \quad \text{------(2)}$$

When the difference of pressure between two points 1 and 2 by using U-tube manometer as shown in figure, we have

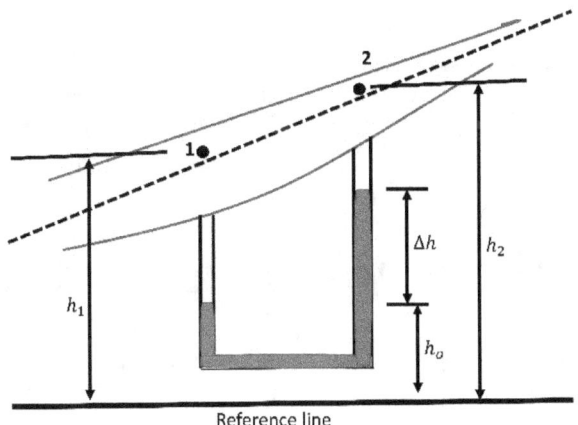

$$P_1+\rho g(h_1-h_o)=P_2+\rho g(h_2-h_0-\Delta h)+\Delta h\rho_m g$$

$$\Rightarrow (P_1+\rho gh_1)-(P_2+\rho gh_2)=(\rho_m-\rho)g\Delta h$$

$$\Rightarrow \left[\dfrac{P_1-P_2}{\rho}-g(h_2-h_1)\right]=\left[\dfrac{\rho_m}{\rho}-1\right]\Delta h$$

Thus, equ (4) becomes

$$Q_2=\dfrac{A_1A_2}{\sqrt{A_1^2-A_2^2}}\sqrt{2g\left[\dfrac{\rho_m}{\rho}-1\right]\Delta h} \quad \text{------(3)}$$

Consider the coefficient of contraction C_c defined by

$$C_c = \frac{A_2}{A_o} \Rightarrow A_2 = C_c A_o$$

Substituting the value of A_2 in equ (3), we obtain as

$$Q_2 = \frac{A_1 C_c A_o}{\sqrt{A_1^2 - (C_c A_o)^2}} \sqrt{2g \left[\frac{\rho_m}{\rho} - 1 \right] \Delta h}$$

$$\Rightarrow Q_2 = C \sqrt{\left[\frac{\rho_m}{\rho} - 1 \right] \Delta h}$$

The value of C depends on the ratio of orifice to the duct area and Reynolds number of flowing fluid. A differential manometer is attached to the section where the vena contracta occurs, but in practice, it has been difficult to get accurate reading. So, determination of accurate values of C of an orificemeter at different operating conditions is known as calibration of the orificemeter.

DEDUCTION: TORRIECLLI'S THEOREM

Consider a cylindrical tank filled with water. It is completely sealed except a small orifice close to its base. Let A_1 and A_2 are the cross-sectional areas of the tank and of the orifice respectively. Let us consider that a water jet exit from the orifice in the streamline connecting two points 1 and 2 as shown in the figure.

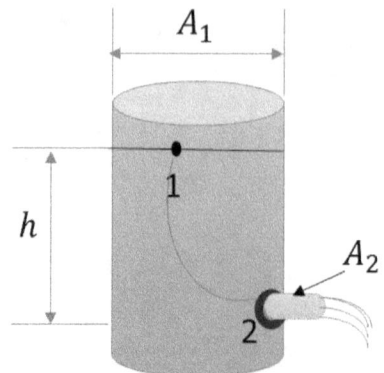

So we can apply the Bernoulli's theorem along the streamline from the point 1 to the point 2, we obtain as:

$$\frac{P_1}{\rho g} + \frac{1}{2}v_1^2 + h_1 = \frac{P_2}{\rho g} + \frac{1}{2}v_2^2 + h_2$$

$$\Rightarrow \frac{1}{2}(v_2^2 - v_1^2) = \frac{1}{\rho}(P_1 - P_2) + g(h_1 - h_2) ----(1)$$

According to the principle of continuity, we have:

$$v_1 A_1 = v_2 A_2$$

$$\Rightarrow v_1 = \left(\frac{A_2}{A_1}\right) v_2$$

Now the equ (1) becomes

$$\Rightarrow \frac{1}{2}\left[v_2^2 - v_2^2\left(\frac{A_2}{A_1}\right)^2\right] = \frac{1}{\rho}(P_1 - P_2) + g(h_1 - h_2)$$

$$\Rightarrow \frac{1}{2}v_2^2\left[1 - \left(\frac{A_2}{A_1}\right)^2\right] = \frac{1}{\rho}(P_1 - P_2) + g(h_1 - h_2) --(2)$$

From the figure, we have

$$h_1 - h_2 = h$$

Now the equ (2) becomes

$$\Rightarrow \frac{1}{2}v_2^2\left[1 - \left(\frac{A_2}{A_1}\right)^2\right] = \frac{1}{\rho}(P_1 - P_2) + gh$$

$$\Rightarrow v_2^2 = \frac{2}{\left[1 - \left(\frac{A_2}{A_1}\right)^2\right]}\left[\frac{P_1 - P_2}{\rho} + gh\right]$$

$$\Rightarrow v_2 = \sqrt{\frac{2}{\left[1 - \left(\frac{A_2}{A_1}\right)^2\right]}\left[\frac{P_1 - P_2}{\rho} + gh\right]} ----(3)$$

This expression indicated the speed of efflux of water jet from the orifice.

Here, if upper surface of cylindrical tank would be open then we can say that pressure of the fluid in the tank and pressure of the efflux water from the orifice both are same to atmospheric pressure.

$$P_1 = P_2 = P_{atm}$$

From the figure we observe that

$$A_2 <<< A_1$$

$$\Rightarrow \frac{A_2}{A_1} <<< 1$$

$$\Rightarrow 1 - \left(\frac{A_2}{A_1}\right)^2 <<< 0$$

Then, the equ (3) becomes

$$v_2 = \sqrt{2gh}$$

The above expression determine the speed of efflux of water jet from the orifice which is equivalent to the expression for the rigid body freely falling from a height h. So, we can states that the velocity water jet exit from orifice is proportional to the square root of the height above the orifice, is known as **Torrieclli's Theorem**.

Moreover, the expression of speed of efflux is calculated on neglecting the frictional forces.

So, in order to obtain the expression of actual speed of efflux of water jet, we need to multiplying one factor called **coefficient of velocity** , C_v Then speed of efflux is

$$v_{actual} = C_v \sqrt{2gh}$$

Next, the actual volumetric flow rate is given by

$$Q = A_2 v_{actual} = A_2 C_v \sqrt{2gh}$$

Where A_2 be the cross-sectional area of water jet at Ven-contracta.

Consider the **coefficient of contraction** , C_c defined by

$$C_c = \frac{A_2}{A_o} \Rightarrow A_2 = C_c A_o$$

Then, the volumetric flow rate is given by

$$Q = A_2 v_{actual} = A_2 C_v \sqrt{2gh} = C_c A_o C_v \sqrt{2gh} \quad ----(4)$$

Moreover we recall that **coefficient of discharge** is define:

$$C_d = C_c C_v \text{ Hence, the equ (4) becomes}$$

$$Q = C_d A_o \sqrt{2gh}$$

APLLICATIONS: Few major applications as mention below:
1. It is used in Engine Carburetors to measure airflow relevant in automobile industry.
2. It is commonly used in process industries to control flow rate.
3. It is also used in medical industry as blood flows in the arteries measured by this.
4. The flow rate of oil and gasoline taken by venturimeters in Oil & Gas Industries.

ADVANTAGES: The Orificemeter is widely used in many industries because of following advantages as
1. The foremost advantage of this is very cheaply available as compared to the venturimeter.
2. The space required for installation is less.
3. This system provides very less pressure drop.
4. The system is capable to used for variety of fluids.
5. Its maintainance cost is very low.
6. The system can be placed in refered positions whether it can be vertical, horizontal or inclined at certain angle.
7. The orifice plate is so thin that it fitted in between the existing pipe.

DISADVANTAGES: However, there are certain limitations such as:
1. Due to low cross sectional area of orifice plate, sometimes it is difficult to read the pressure drop.
2. Here, downstream pressure can not be measured.
3. The accuracy in reading by Orificemeter is greatly affected by viscosity, density, and pressure of the fluid.
4. Here, the coefficient of discharge is very low as compared to the venturimeter.
5. The system is suitable in straight pipe usually.

SOLVED EXAMPLES

EXAMPLE 1. An orifice meter with orifice diameter 15 cm is inserted in a pipe of 30 cm diameter as shown in figure. The pressure gauges fitted upstream and downstream of the orifice meter gives readings of 14.715 N/cm^2 and 9.81 N/cm^2 respectively. Determine the discharge rate of flow of water through the pipe provided the coefficient of discharge through orificemeter is 0.6

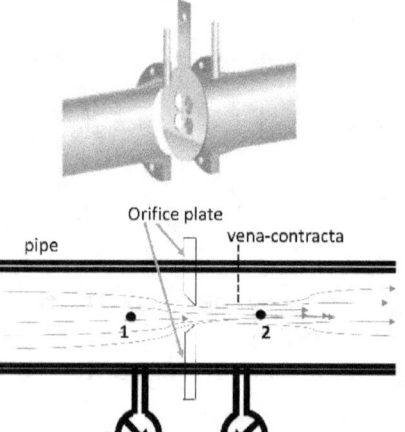

Orifice plate

pipe   vena-contracta

Pressure gauge

Answer: In this case,

d_o = diamter of orifice = 15 cm

A_1 = cross-sectional area of orifice plate

$$= \frac{\pi}{4}d_o^2 = \frac{\pi}{4}(15)^2 = 176.625 \text{ cm}^2$$

d_p = diamter of pipe line = 30 cm

A_2 = cross-sectional area of ven-contracta portion of pipe line

$$= \frac{\pi}{4}d_p^2 = \frac{\pi}{4}(30)^2 = 706.5 \text{ cm}^2$$

C_d = coefficient of discharge through orificemeter = 0.6

P_1 = pressure at point 1 = 14.715 N/cm² = 14.715×10^4 N/m²

$\therefore \quad \dfrac{P_1}{\rho g}$ = pressure head at the point 1 = $\dfrac{14.715 \times 10^4}{1000 \times 9.81} = 15$ m

P_2 = pressure at point 2 = 9.81 N/cm² = 9.81×10^4 N/m²

$\therefore \quad \dfrac{P_2}{\rho g}$ = pressure head at the point 2 = $\dfrac{9.81 \times 10^4}{1000 \times 9.81} = 10$ m

$\therefore \quad h = \dfrac{P_1}{\rho g} - \dfrac{P_2}{\rho g} = 15 - 10 = 5$ m = 500 cm

The discharge through orifice meter is

98

$$Q = C_d \frac{A_1 A_2}{\sqrt{A_2^2 - A_1^2}} \sqrt{2gh}$$

$$\Rightarrow Q = 0.6 \frac{176.625 \times 706.5}{\sqrt{(706.5)^2 - (176.625)^2}} \sqrt{2 \times 9.81 \times 500}$$

$$\Rightarrow Q = 0.6 \times 182.4 \times 99.045$$

$$\Rightarrow Q = 10839.49 \text{ cm}^3 / \sec$$

$$\Rightarrow Q = 10.83 \text{ Lt/sec}$$

EXAMPLE 2. An orifice meter with orifice diameter 15 cm is inserted in a pipe of 30 cm diameter as shown in figure. The pressure gauges fitted upstream and downstream of the orifice meter gives readings of 14.715 N/cm^2 and 9.81 N/cm^2 respectively. Determine the discharge rate of flow of oil of density 800 kg/m^3 through the pipe provided the coefficient of discharge through orificemeter is 0.6.

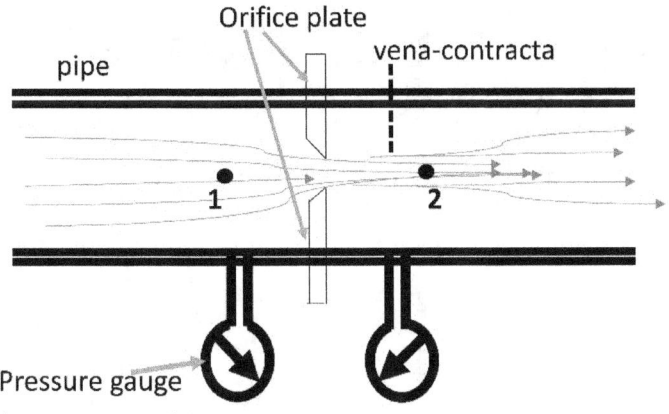

Answer: In this case,

d_o = diamter of orifice = 15 cm

A_1 = cross-sectional area of orifice plate

$$= \frac{\pi}{4} d_o^2 = \frac{\pi}{4}(15)^2 = 176.625 \text{ cm}^2$$

d_p = diamter of pipe line = 30 cm

A_2 = cross-sectional area of ven-contracta portion of pipe line

$$= \frac{\pi}{4} d_p^2 = \frac{\pi}{4} (30)^2 = 706.5 \text{ cm}^2$$

C_d = coefficient of discharge through orificemeter = 0.6

ρ = density of oil = 800 kg/m³

P_1 = pressure at point 1 = 14.715 N/cm² = 14.715×10⁴ N/m²

$\therefore \quad \dfrac{P_1}{\rho g}$ = pressure head at the point 1 $= \dfrac{14.715 \times 10^4}{800 \times 9.81} = 18.75$ m

P_2 = pressure at point 2 = 9.81 N/cm² = 9.81×10⁴ N/m²

$\therefore \quad \dfrac{P_2}{\rho g}$ = pressure head at the point 2 $= \dfrac{9.81 \times 10^4}{800 \times 9.81} = 12.5$ m

$\therefore \quad h = \dfrac{P_1}{\rho g} - \dfrac{P_2}{\rho g} = 18.75 - 12.5 = 6.25$ m = 625 cm

The discharge through orifice meter is

$$Q = C_d \frac{A_1 A_2}{\sqrt{A_2^2 - A_1^2}} \sqrt{2gh}$$

$$\Rightarrow Q = 0.6 \frac{176.625 \times 706.5}{\sqrt{(706.5)^2 - (176.625)^2}} \sqrt{2 \times 9.81 \times 625}$$

$$\Rightarrow Q = 0.6 \times 182.4 \times 110.74$$

$$\Rightarrow Q = 12120 \text{ cm}^3 / \text{sec}$$

$$\Rightarrow Q = 12.12 \text{ Lt/sec}$$

PITOT TUBE

A pitot tube is a device used to measure the fluid motion at the particular point in the stream flow.

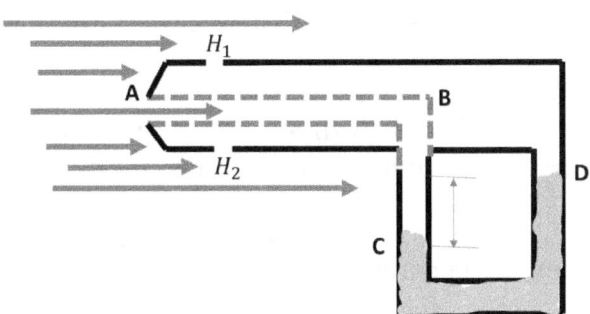

Consider the inner tube BA is kept so as to face the direction of the flow as shown in figure. The outer tube has few holes such as H_1 and H_2. Let P be the pressure in the stream flow such that it remain same inside and outside of the holes and v be the fluid velocity. Now, assume that the fluid enters the tube AB and let it level up to the mark C where stagnation pressure holds. Let P_o be the pressure at point C.

Next, applying the Bernoulli's equation along the streamline from the point A and point C, we have

$$\frac{P_A}{\rho g} + \frac{v_A^2}{2g} + h_A = \frac{P_C}{\rho g} + \frac{v_C^2}{2g} + h_C ---(1)$$

Since there is no elevation of heights from the zero point of reference, so can write as:

$$h_A = h_C$$

Again recall that mercury fluid inside the tube is in stationary so we can say that

$$v_C = 0$$

Then, the equ (1) becomes

$$\frac{P}{\rho g} + \frac{v^2}{2g} = \frac{P_o}{\rho g}$$

$$\Rightarrow \frac{P}{\rho} + \frac{v^2}{2} = \frac{P_o}{\rho}$$

$$\Rightarrow \frac{1}{2}v^2 = \frac{P_o - P}{\rho}$$

$$\Rightarrow v = \sqrt{\frac{2(P_o - P)}{\rho}}$$

Let h be column height in differential U-tube manometer and ρ_m be the density of mercury inside the U-tube. We have

$$P - P_o = \rho_m g h$$

Thus, the equ (2) becomes

$$v = \sqrt{\frac{2\rho_m g h}{\rho}}$$

$$\Rightarrow v = C_v \sqrt{2gh} \qquad \left[C_v = \frac{\rho_m}{\rho} = \text{coefficient of pitot tube} \right]$$

The above expression gives the fluid velocity at a point in streamline of fluid flow through pitot tube.

SOLVED EXAMPLES

EXAMPLE 1. The pressure difference measured by the two tappings of a pitot-static tube, one tapping pointing upstream and other perpendicular to the flow, placed in the centre of a pipe line of diameter 40 cm is 10 cm of water as shown in figure. The mean velocity in the pipe is 0.75 times the central velocity. Determine the discharge through the pipe provided the coefficient of discharge through pitot tube is 0.98.

Answer: In this case,

d = diameter of pipe line = 40 cm

h = difference of pressure head = 10 cm

C_v = coefficient of pitot tube = 0.98

Central velocity at a point in streamline of fluid flow through pitot tube is given by

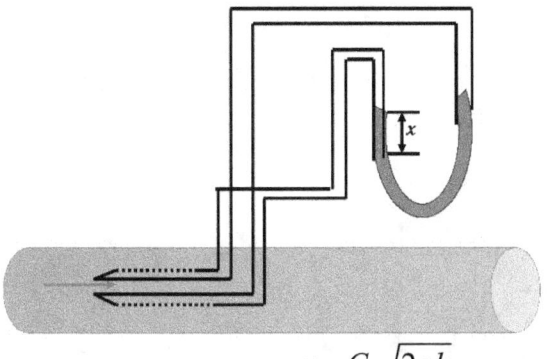

$$v = C_v\sqrt{2gh}$$

$$\Rightarrow v = 0.98 \times \sqrt{2 \times 9.81 \times 10}$$

$$\Rightarrow v = 0.98 \times \sqrt{196.2}$$

$$\Rightarrow v = 0.98 \times 14.007$$

$$\Rightarrow v = 13.727 \text{ cm/sec}$$

V = mean velocity = $0.75 \times$ central velocity

$$\Rightarrow V = 0.75 \times 13.727$$

$$\Rightarrow V = 10.30 \text{ cm/sec}$$

Discharge through pipe is

$$Q = \text{Area of pipe} \times V$$

$$\Rightarrow Q = \frac{\pi}{4}d^2 \times V$$

$$\Rightarrow Q = \frac{\pi}{4}(40)^2 \times 10.30$$

$$\Rightarrow Q = 12936.8 \text{ cm}^3 / \text{sec}$$

$$\Rightarrow Q = 12.9368 \text{ Lt/sec}$$

EXAMPLE 2. Determine the velocity of flow of an oil through a pipe, when the difference of mercury level in a differential U-tube manometer connected to the two tappings of the pitot-tube is 15 cm as shown in figure. Here, assume the coefficient of discharge through pitot tube is 0.98 and density of oil is 800 kg / m^3.

Answer: In this case,

y = difference of mercury level in differential U-tube $= 15$ cm

ρ_m = density of mercury $= 13545.848$ kg/m^3

ρ_o = density of oil $= 800$ kg/m^3

C_v = coefficient of discharge through pitot tube $= 0.98$

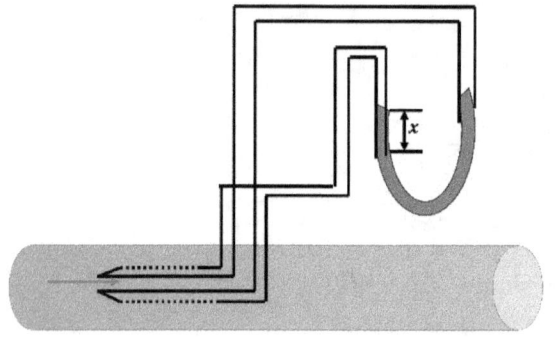

$\therefore$ Difference of pressure heads $h = y\left[\dfrac{\rho_m}{\rho_o} - 1\right]$

$$\Rightarrow h = 15\left[\frac{13545.848}{800} - 1\right]$$

$$\Rightarrow h = 15(16) = 240 \text{ cm}$$

The velocity of flow of an oil through a pipe is

$$v = C_v\sqrt{2gh}$$

$$\Rightarrow v = 0.98 \times \sqrt{2 \times 9.81 \times 240}$$

$$\Rightarrow v = 0.98 \times \sqrt{4708.8}$$

$$\Rightarrow v = 0.98 \times 68.621$$

$$\Rightarrow v = 67.24858 \text{ cm/sec}$$

EXAMPLE 3. A submarine moves horizontally in sea and its axis 20 m below the surface of water. A pitot-static tube placed in front of sub-marine and along its axis, is connected to the two limbs of a U-tube containing mercury as shown in figure. The difference of mercury level is found to be 20 cm. Determine the speed of submarine if density of sea water is 1026 kg/m^3.

Answer: In this case,

ρ_m = density of mercury $= 13545.848$ kg/m^3

ρ_w = density of sea water $= 1026 \text{ kg/m}^3$

y = difference of mercury level $= 20 \text{ cm}$

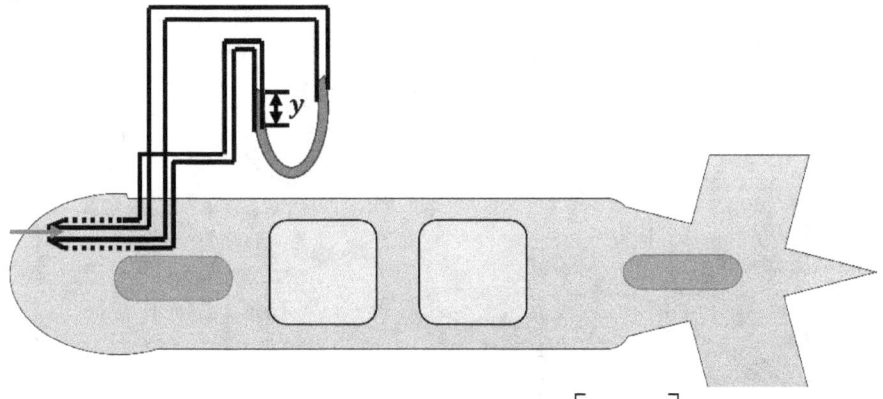

$\therefore$ Difference of pressure heads, $h = y\left[\dfrac{\rho_m}{\rho_w} - 1\right]$

$$\Rightarrow h = 20\left[\frac{13545.848}{1026} - 1\right]$$

$$\Rightarrow h = 20(12.20) = 244 \text{ cm}$$

The speed of submarine is $v = C_d\sqrt{2gh}$

$$\Rightarrow v = 0.98\sqrt{2 \times 9.81 \times 244}$$

$$\Rightarrow v = 0.98 \times 69.20$$

$$\Rightarrow v = 67.816 \text{ cm/sec}$$

EXAMPLE 4. Determine the velocity of flow of water in a pipe of diameter 300 mm at a point, where the stagnation pressure head is 5 m and static pressure head is 4 m provided the coefficient of the pitot-tube is 0.97.

Answer: In this case,

h_s = stagnation pressure head $= 5 \text{ m}$

h_t = static pressure head $= 4 \text{ m}$

$\therefore$ Difference in pressure head, $h = h_s - h_t = 5 - 4 = 1 \text{ m}$

The velocity of water flowing through pipe is

$$v = C_d\sqrt{2gh}$$
$$\Rightarrow v = 0.98 \times \sqrt{2 \times 9.81 \times 1}$$
$$\Rightarrow v = 4.34 \text{ m/s}$$

NOZZLES

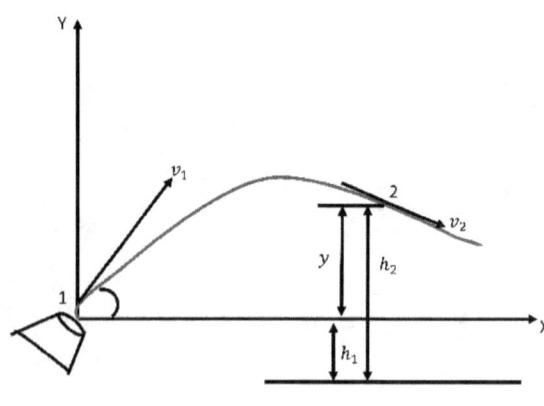

Consider a fluid exit from a nozzle of cross sectional area A in the form of jet as shown in figure. The fluid jet is moving with the velocity v_1 and making a angle α with the horizon. Let us assume that fluid jet is in the atmosphere so that $P_1 = P_2$.

Now, applying applying the Bernoulli's equation along the streamline from the point 1 to point 2, we obtain as:

$$\frac{P_1}{\rho g} + \frac{v_1^2}{2g} + h_1 = \frac{P_2}{\rho g} + \frac{v_2^2}{2g} + h_2$$
$$\Rightarrow \frac{v_1^2}{2g} + h_1 = \frac{v_2^2}{2g} + h_2$$
$$\Rightarrow v_2^2 = v_1^2 - 2g(h_2 - h_1)$$
$$\Rightarrow v_2^2 = v_1^2 - 2gy ----(1)$$

Then, the volumetric flow rate of fluid is define as

$$Q = v_1 A \Rightarrow v_1 = \frac{Q}{A}$$

Then the equ (1) becomes

$$v_2^2 = \left(\frac{Q}{A}\right)^2 - 2gy ----(2)$$

To determine the trajectory of the fluid jet, consider the equations of the motion for the jet along the horizontal line (x-axis) and vertical line (y-axis) where (x,y) are coordinates of the point 2 and point 1. Then we have

$$\frac{dx}{dt} = v_1 \cos \alpha \ - - - - (3)$$

$$\frac{dy}{dt} = v_1 \sin \alpha - gt - - - - (4) \left[\because \cos \alpha = \frac{B}{H} = \frac{\frac{dx}{dt}}{v_1} \right]$$

Integrating (3) and (4), we get

$$x = (v_1 \cos \alpha)t + c_1 \ - - - (5)$$

$$y = (v_1 \sin \alpha)t - \frac{1}{2}gt^2 + c_2 \ - - - - (6)$$

Initially, at the point 1, $t = 0, x = 0, y = 0$

$$\Rightarrow c_1 = 0 \quad and \quad c_2 = 0$$

Then, the equ (5) and (6) becomes as

$$x = (v_1 \cos \alpha)t \ - - - - - (7)$$

$$y = (v_1 \sin \alpha)t - \frac{1}{2}gt^2 \ - - - - (8)$$

(7) $\times \sin \alpha \Rightarrow x \sin \alpha = v_1 \sin \alpha \cos \alpha.t$

(8) $\times \cos \alpha \Rightarrow y \cos \alpha = v_1 \sin \alpha \cos \alpha.t - \frac{1}{2}gt^2 \cos \alpha$

$$\Rightarrow y \cos \alpha - x \sin \alpha = -\frac{1}{2}gt^2 \cos \alpha$$

$$\Rightarrow y \cos \alpha = x \sin \alpha - \frac{1}{2}g\,t^2 \cos \alpha \ - - - - - (9)$$

From (7), we get

$$x = (v_1 \cos \alpha).t$$

$$\Rightarrow t = \frac{x}{v_1 \cos \alpha} = \frac{x}{v_1} \sec \alpha$$

Next the equ (9) reduced to

$$y = x \tan \alpha - \frac{1}{2}g\frac{x^2}{v_1^2}\sec^2 \alpha \ - - - - - (10)$$

Recall that

$$v_1 = \frac{Q}{A} \Rightarrow \frac{1}{v_1} = \frac{A}{Q} \Rightarrow \frac{1}{v_1^2} = \left(\frac{A}{Q}\right)^2$$

The equ (10) becomes

$$y = x \tan \alpha - \frac{x^2}{2}g\left(\frac{A}{Q}\right)^2 \sec^2 \alpha$$

Putting this value of y in the equ (2), the velocity of the fluid jet at any point is define as

107

$$v_2^2 = \left(\frac{Q}{A}\right)^2 - 2g\,x\tan\alpha + x^2 g^2\left(\frac{A}{Q}\right)^2 \sec^2\alpha$$

SOLVED EXAMPLES

EXAMPLE 1. Determine the fluid motion for the jet of water squeezed from a nozzle which is inclined at an angle of 60° along with situated at a distance of 1.2 m above the ground level. The jet of water strikes a distance of 5 m on the horizontal griund level.

Solution: As per detail discussion in this chapter, we have the formulae to calculate the horizontal distance as:

$$y = x\tan\alpha - \frac{g\,x^2}{2\,v^2\cos\alpha}$$

Given that $y = -1.2\,m,\; x = 5\,m,\; \alpha = 60^o$

$$\Rightarrow -1.2\,m = 5\tan 60^o - \frac{\left(9.81\frac{m}{s^2}\right)(5^2\,m^2)}{2\,v^2}\sec^2 60^o$$

$$\Rightarrow v = 7.05\,\frac{m}{s}$$

EXAMPLE 2. Determine the horizontal distance covered by the jet of water squeezed from a nozzle which is inclined at an angle of 30° along with situated at a distance of 2 m above the ground level. The water discharge through the nozzle at the rate of 25 m/s.

Solution: As per detail discussion in this chapter, we have the formulae to calculate the horizontal distance as:

$$y = x\tan\alpha - \frac{g\,x^2}{2\,v^2\cos\alpha}$$

Given that $y = -2\,m,\; v = 25\frac{m}{s},\; \alpha = 30^o$

$$\Rightarrow -2\,m = x\tan 30^o - \frac{1}{2}\left(\frac{9.81\frac{m}{s^2}}{25^2\frac{m^2}{s^2}}\right)\sec^2 30^o$$

$$\Rightarrow x^2 - 55.55\,x - 192.3 = 0$$

$$\Rightarrow x = 58.82\,m$$

EXAMPLE 3. A nozzle of diameter 30 mm is fitted to a pipe of 60 mm diameter as shown in figure. Determine the force exerted by the nozzle on the water which is flowing through the pipe at the rate of 4 m^3 / minute.

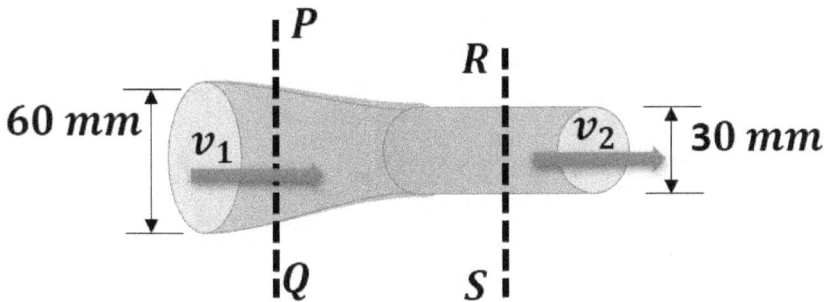

Answer: In this case,

d_1 = diameter of pipe = 60 mm = 60×10^{-3} m = 0.06 m

a_1 = cross-sectional aea of pipe line

$$= \frac{\pi}{4} d_1^2 = \frac{\pi}{4} (0.06)^2 = 0.002826 \text{ m}^2$$

d_2 = diamter of nozzle = 30 mm = 30×10^{-3} m = 0.03 m

a_2 = cross-sectional arera of nozzle

$$= \frac{\pi}{4} d_2^2 = \frac{\pi}{4} (0.03)^2 = 0.0007065 \text{ m}^2$$

Q = discharge = 4 m³ / min = $\dfrac{4}{60}$ m³ / sec = 0.067 m³ / sec

Applying continuity equation at the section PQ and RS,

$$a_1 v_1 = a_2 v_2 = Q$$

∴

$$v_1 = \frac{Q}{a_1} = \frac{0.067}{0.002826} = 23.71 \text{ m/s}$$

$$v_2 = \frac{Q}{a_2} = \frac{0.067}{0.0007065} = 94.83 \text{ m/s}$$

Next, apply Bernoulli's equation at the section PQ and RS,

$$\frac{P_1}{\rho g} + \frac{v_1^2}{2g} + z_1 = \frac{P_2}{\rho g} + \frac{v_2^2}{2g} + z_2 \quad \text{------(1)}$$

Since, nozzle is installed in horizontal pipe line, then we have

$$z_1 = z_2$$

Since, outlet of nozzle is open to atmosphere, then assume

$$P_2 = 0$$

$$\Rightarrow \frac{P_2}{\rho g} = 0$$

Next, the equation (1) reduced to

$$\frac{P_1}{\rho g} + \frac{v_1^2}{2g} = \frac{v_2^2}{2g}$$

$$\Rightarrow \frac{P_1}{\rho g} = \frac{v_2^2}{2g} - \frac{v_1^2}{2g}$$

$$\Rightarrow \frac{P_1}{\rho g} = \frac{(94.83)^2}{2 \times 9.81} - \frac{(23.71)^2}{2 \times 9.81}$$

$$\Rightarrow \frac{P_1}{\rho g} = \frac{8992.7289}{19.62} - \frac{562.1641}{19.62}$$

$$\Rightarrow \frac{P_1}{\rho g} = 458.345 - 28.65$$

$$\Rightarrow \frac{P_1}{\rho g} = 429.695 \text{ m of water}$$

$$\Rightarrow P_1 = 429.695 \times 1000 \times 9.81$$

$$\Rightarrow P_1 = 4215307.95 \text{ N/m}^2$$

Let the force exerted by the nozzle on water $= F_x$

Net force in $x-$ direction $=$ rate of change of momentum in $x-$ direction

∴

$$P_1 a_1 - P_2 a_2 + F_x = \rho Q(v_2 - v_1)$$

$$\Rightarrow 4215307.95 \times 0.002826 - 0 + F_x = 1000 \times 0.067(94.83 - 23.71)$$
$$\Rightarrow 11912.46 + F_x = 697687.2$$
$$\Rightarrow F_x = 697687.2 - 11912.46$$
$$\Rightarrow F_x = 685774.74 \text{ N/m}^2$$

EXAMPLE 4. A lawn sprinkler with two nozzles of diameters 3 mm each is connected across a tap of water as shown in figure. The nozzles are at a distance of 40 cm and 30 cm from the centre of the tap. The rate of water through tap is 100 cm^3/sec. The nozzle discharges water in the downward directions. Determine the angular speed at which the sprinkler will rotate free.

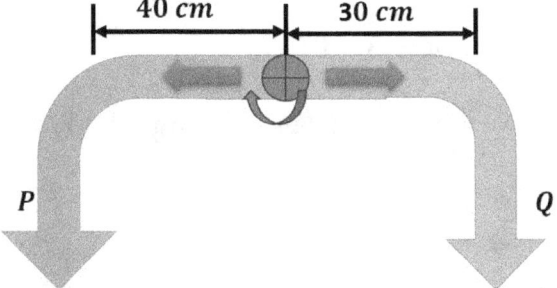

Answer: In this case,

$d = d_P = d_Q$ = diameter of nozzles at P and Q

$= 3$ mm $= 0.003$ m

$\therefore$ A = cross-sectional area of pipe of sprinkler

$$\Rightarrow A = \frac{\pi}{4}(0.003)^2 = 7.065 \times 10^{-6} \text{ m}^2$$

Q = discharge rate of water flows

$= 100$ cm^3/sec $= 100 \times 10^{-6}$ cm^3/sec

Consider that discharge to be equally divided between the two nozzles, we have

$$Q_P = Q_Q = \frac{Q}{2} = \frac{100 \times 10^{-6}}{2} = 50 \times 10^{-6} \text{ m}^3 \text{/sec}$$

$\therefore$ speed of water at the oulet of each nozzle,

$$v_P = v_Q = \frac{Q}{A} = \frac{50 \times 10^{-6}}{7.065 \times 10^{-6}} = 7.077 \text{ m/s}$$

The jet of water coming out from nozzles P and Q is having speed 7.077 m/s. These jets of water will exert force in the upward direction. The torque exerted will also be in the opposite direction. Hence, torque at Q will be in the anti-clockwise direction and at P in the clockwise direction. But torque at Q is more than the torque at P and hence sprinkle, if free it will rotate in the anti-clockwise direction as shown in figure.

ω = angular speed of the lawn sprinkler

Then absolute speed of water flows from point P is

$$v_1 = v_P + \text{tangential speed due to rotation}$$
$$\Rightarrow v_1 = v_P + \omega \times r_P$$
$$\Rightarrow v_1 = (7.077 + \omega \times 0.3) \text{ m/s}$$

Similarly, absolute speed of water flows from the point Q is

$$v_2 = v_Q - \text{tangential speed due to rotation}$$
$$\Rightarrow v_2 = 7.077 - \omega \times r_Q$$
$$\Rightarrow v_2 = (7.077 - \omega \times 0.4) \text{ m/s}$$

Next, according to moment of momentum principle,

Torque = rate of change of moment of momentum

$$\Rightarrow T = \rho Q \left(v_2 r_Q - v_1 r_P \right)$$

$$\Rightarrow T = 1000 \times 100 \times 10^{-6} \left[(7.077 - 0.4\omega) \times 0.4 - (7.077 + 0.3\omega) \times 0.3 \right]$$

Since there is no external torque applied on sprinkler. Hence torque is zero, i.e., $T = 0$

$$\therefore 1000 \times 100 \times 10^{-6} \left[(7.077 - 0.4\omega) \times 0.4 - (7.077 + 0.3\omega) \times 0.3 \right] = 0$$

$$\Rightarrow \left[(7.077 - 0.4\omega) \times 0.4 - (7.077 + 0.3\omega) \times 0.3 \right] = 0$$

$$\Rightarrow 7.077 \times 0.4 - 0.16\omega - 7.077 \times 0.3 - 0.09\omega = 0$$

$$\Rightarrow 7.077 \times (0.4 - 0.3) + \omega(-0.16 - 0.09) = 0$$
$$\Rightarrow 7.077 \times 0.1 + \omega(-0.25) = 0$$
$$\Rightarrow 0.7077 = 0.25\omega$$
$$\Rightarrow \omega = \frac{0.7077}{0.25} = 2.8308 \text{ rad/sec}$$

EXAMPLE 5. A lawn sprinkler has two nozzles of diameters 8 mm each at the end of a rotating arm and the velocity of flow of water from each nozzle is 12 $m/\sec$. On nozzle discharges water in the downward direction, while the other nozzle discharges water vertically up as shown in figure. The nozzles are at a distance of 40 cm from the centre of the rotating arm. Determine the speed of rotation of the arm, if it is free to rotate.

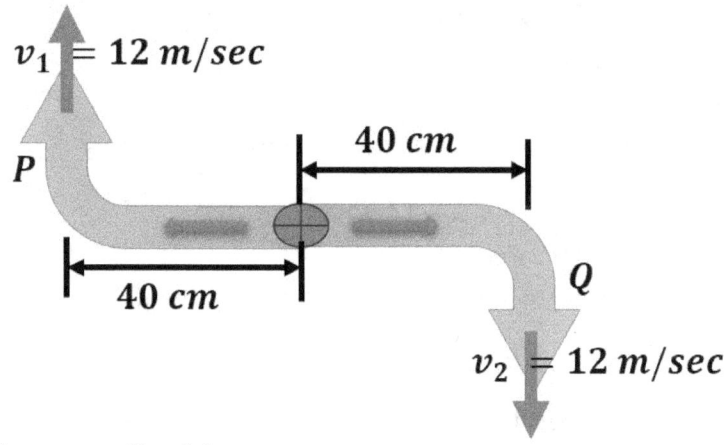

Answer: In this case,

d = diameter of each nozzle = 8 mm = 0.008 m

∴ A = cross-sectional area of each nozzle = $\dfrac{\pi}{4}d^2$

$$\Rightarrow A = \frac{\pi}{4}(0.008)^2 = 5.026 \times 10^{-5} \text{ m}^2$$

v = speed of water flows through each nozzle = 12 m/s

ω = speed of rotation of the sprinkler

$r = r_P = r_Q = $ distance of nozzles P and Q from the centre of tap

$= 40$ cm $= 0.4$ m

$\therefore \quad Q = $ Discharge of water through each nozzle

$$\Rightarrow Q = Av$$
$$\Rightarrow Q = 5.026 \times 10^{-5} \times 12$$
$$\Rightarrow Q = 60.312 \times 10^{-5} \text{ m}^3 / \sec$$

The absolute speed of water flows through nozzle P is
$$v_1 = 12 - 0.4 \times \omega$$

Torque exerted by water coming out from nozzle P is

$$= r_P \times \rho \times Q \times v_1$$
$$= 0.4 \times 1000 \times 60.312 \times 10^{-5} \times (12 - 0.4\omega)$$
$$= 0.241248(12 - 0.4\omega)$$

The absolute speed of water flows through nozzle Q is

$$v_2 = 12 - 0.4\omega$$

Torque exerted by water coming out from nozzle Q is

$$= r_Q \times \rho \times Q \times v_2$$
$$= 0.4 \times 1000 \times 60.312 \times 10^{-5} \times (12 - 0.4\omega)$$
$$= 0.241248(12 - 0.4\omega)$$

Total torque exerted by water coming out from nozzles is

$$= 0.241248(12 - 0.4\omega) + 0.241248(12 - 0.4\omega)$$

Since there is no external torque is applied on sprinkler, so total torque on the sprinkler must be zero.

$$\therefore \quad 0.241248(12-0.4\omega)+0.241248(12-0.4\omega)=0$$

$$\Rightarrow 0.241248\times24-0.241248\times0.8\omega=0$$

$$\Rightarrow 5.789952-0.1929984\omega=0$$

$$\Rightarrow \omega=\frac{5.789952}{0.1929984}$$

$$\Rightarrow \omega=30 \text{ rad/sec}$$

$$\Rightarrow k=\frac{60\times\omega}{2\pi}=\frac{60\times30}{2\times3.14}=286.62 \text{ r.p.m}$$

EXAMPLE 6. A vertical wall is of 10 m in height. A jet of water is issuing from a nozzle with a velocity of 25 m/s. The nozzle is situated at a horizontal distance of 20 m from the vertical wall as shown in figure. Determine the angle of projection of the nozzle to the horizontal so that the jet of water just clears the tap of the wall.

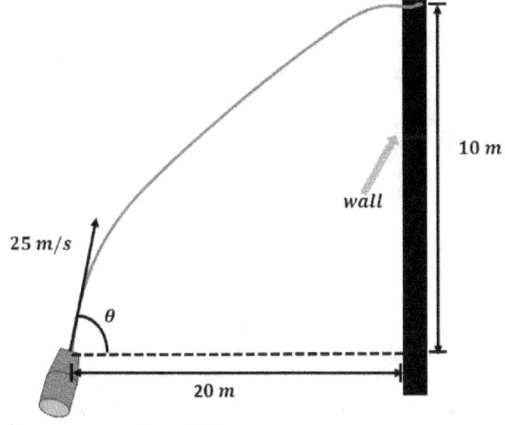

Answer: In this case,

$y=$ height of vertical wall $=10$ m

$v=$ velocity of water jet coming out of nozzle $=25$ m/s

$x=$ distance of water jet from the wall $=20$ m

Let α be the required angle of projection of nozzle to the horizontal so that jet of water clears the tap of wall

The height of wall from the jet of water is defined as

$$y = x \tan \alpha - \frac{x^2}{2v^2} g \sec^2 \alpha$$

$$\Rightarrow 10 = 20 \tan \alpha - \frac{(20)^2}{2(25)^2} \times 9.81 \times (1 + \tan^2 \alpha)$$

$$\Rightarrow 10 = 20 \tan \alpha - 3.1392 \times (1 + \tan^2 \alpha)$$

$$\Rightarrow 3.1392 \tan^2 \alpha - 20 \tan \alpha + 13.1392 = 0$$

$$\Rightarrow \tan \alpha = \frac{20 \pm \sqrt{(20)^2 - 4 \times 3.1392 \times 13.1392}}{2 \times 3.1392}$$

$$\Rightarrow \tan \alpha = \frac{20 \pm \sqrt{400 - 164.99}}{6.2784}$$

$$\Rightarrow \tan \alpha = \frac{20 \pm \sqrt{235.01}}{6.2784}$$

$$\Rightarrow \tan \alpha = \frac{20 \pm 15.33}{6.2784}$$

$$\Rightarrow \tan \alpha = \frac{35.33}{6.2784} \text{ or } \frac{4.67}{6.2784}$$

$$\Rightarrow \tan \alpha = 5.63 \text{ or } 0.74$$

$$\Rightarrow \alpha = 1.40 \text{ rad or } 0.64 \text{ rad}$$

$$\Rightarrow \alpha = 80^0 22' \text{ or } 36^0 67'$$

EXAMPLE 7. A fire-brigade man is holding a fire stream nozzle of 50 mm diameter at a distance of 1 m above the ground and 6 m from a vertical wall as shown in figure. The jet is coming out with a velocity of 15 m/s. This jet is to strike a window, situated at a distance of 11 m above ground in the vertical wall. Determine the angle of inclination with the horizontal made by the jet, coming out from the nozzle. Determine the amount of water falling on the window.

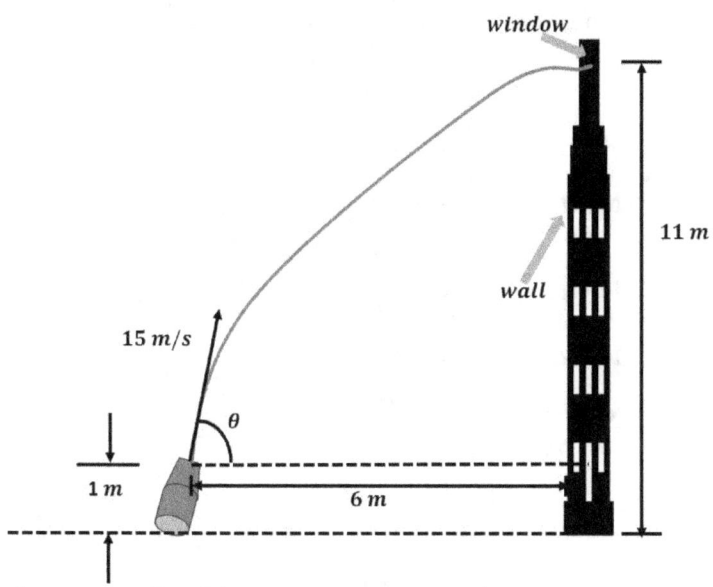

Answer: In this case,

d = diamter of nozzle = 50 mm = 0.05 m

y = height of vertical wall = 11−1 = 10 m

v = velocity of water jet coming out of nozzle = 15 m/s

x = distance of water jet from the wall = 6 m

Let α be the required angle of projection of nozzle to the horizontal so that jet of water strikes the window.

The height of wall from the jet of water is defined as

$$y = x \tan \alpha - \frac{x^2}{2v^2} g \sec^2 \alpha$$

$$10 = 6 \tan \alpha - \frac{6 \times 9.81}{2 \times (15)^2} \sec^2 \alpha$$

$$\Rightarrow 10 = 6 \tan \alpha - \frac{58.86}{450} \sec^2 \alpha$$

$$\Rightarrow 10 = 6 \tan \alpha - 0.1308 \left(1 + \tan^2 \alpha\right)$$

$$\Rightarrow 0.1308 \tan^2 \alpha - 6 \tan \alpha + 10.1308 = 0$$

$$\Rightarrow \tan \alpha = \frac{6 \pm \sqrt{6^2 - 4 \times 0.1308 \times 10.1308}}{2 \times 0.1308}$$

$$\Rightarrow \tan \alpha = \frac{6 \pm \sqrt{36 - 5.300}}{0.2616}$$

$$\Rightarrow \tan \alpha = \frac{6 \pm 5.540}{0.2616}$$

$$\Rightarrow \tan \alpha = \frac{11.540}{0.2616} \quad \text{or} \quad \frac{0.46}{0.2616}$$

$$\Rightarrow \tan \alpha = 44.1131 \text{ or } 1.76$$

$$\Rightarrow \alpha = 1.55 \text{ rad or } 1.05 \text{ rad}$$

$$\Rightarrow \alpha = 88^0 80' \text{ or } 60^0 16'$$

Amount of water falling on the window is

= Discharge of water from the nozzle

= cross-sectional area of nozzle × velocity of jet of water comes out from the nozzle

$$= \frac{\pi}{4} d^2 \times 15$$

$$= \frac{\pi}{4} (0.05)^2 \times 15$$

$$= 0.001963 \times 15$$

$$= 0.029445 \text{ m}^3 / \sec$$

EXAMPLE 8. A window, in a vertical wall, is at a distance of 12 m above the ground level. A jet of water , issuing from a nozzle of diameter 50 mm, is to strike the window as shown in figure. The rate of flow of water through the nozzle is 40 *Lt* / sec. the nozzle is situated at a distance of 1 m above ground level. Determine the greatest horizontal distance from the wall of the nozzle so that jet of water strikes the window.

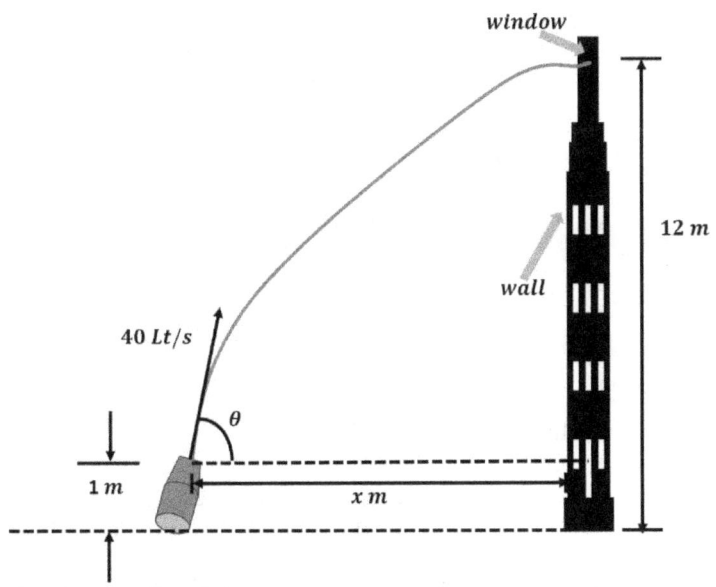

Answer: In this case,

d = diameter of nozzle = 50 mm = 0.05 m

$\therefore$ cross-sectional area of nozzle, $A = \dfrac{\pi}{4}d^2$

$$\Rightarrow A = \frac{\pi}{4}(0.05)^2 = 0.001963 \text{ m}^2$$

Q = discharge = 40 Lt/sec = 40×10^{-3} m³ / sec = 0.04 m³ / sec

Distance of window from ground level = 12 m

Distance of nozzle from ground = 1 m

y = height of vertical window = $12 - 1 = 11$ m

Let x = greatest horizontal distance of the nozzle and

α = angle of inclination to horizontal

v = velocity of water jet coming out of nozzle = $\dfrac{Disch\arg e}{area}$

$$\Rightarrow v = \frac{0.04}{0.001963} = 20.38 \text{ m/s}$$

If the jet reaches the window, then the point Q on the window is on the centre line of the jet. The co-ordinates of Q with respect to P is $(x, 11)$

The height of window from the jet of water is defined as

$$y = x \tan \alpha - \frac{gx^2}{2v^2} \sec^2 \alpha$$

$$\Rightarrow 11 = x \tan \alpha - \frac{9.81x^2}{2 \times (20.38)^2} \sec^2 \alpha$$

$$\Rightarrow 11 = x \tan \alpha - x^2 0.01181 \sec^2 \alpha$$

$$\Rightarrow x \tan \alpha - 0.01181 \frac{x^2}{\cos^2 \alpha} - 11 = 0 \quad \text{------(1)}$$

The maximum value of x with respect to α is obtained by diferentiating the above equation w.r.t. α and put the value of $\frac{dx}{d\alpha} = 0$. Next differentiate the equation (1) w.r.t. α , we obtain as

$$\left[x \sec^2 \alpha + \tan \alpha \times \frac{dx}{d\alpha} \right]$$
$$-0.01181 \left[x^2 \times \left(\frac{(-2)}{\cos^3 \alpha} \right)(-\sin \alpha) + \frac{1}{\cos^2 \alpha} \times 2x \frac{dx}{d\alpha} \right] = 0$$

$$\Rightarrow x \sec^2 \alpha + \tan \alpha \frac{dx}{d\alpha} - 0.01181 \left[\frac{2x^2 \sin \alpha}{\cos^3 \alpha} + \frac{2x}{\cos^3 \alpha} \frac{dx}{d\alpha} \right] = 0$$

For maximum value of x, w.r.t. α, we consider $\frac{dx}{d\alpha} = 0$

Put this value in the above equation, we get

$$x\sec^2\alpha - 0.01181\left[\frac{2x^2\sin\alpha}{\cos^3\alpha}\right] = 0$$

$$\Rightarrow x - 0.02362 \times x^2\frac{\sin\alpha}{\cos\alpha} = 0$$

$$\Rightarrow x\left(1 - 0.02362\tan\alpha \times x\right) = 0$$

$$\Rightarrow 1 - 0.02362\tan\alpha \times x = 0$$

$$\Rightarrow x = \frac{1}{0.02362} \times \frac{1}{\tan\alpha}$$

$$\Rightarrow x = \frac{42.34}{\tan\alpha}$$

Put the value of x in equation (1), we get

$$\frac{42.34}{\tan\alpha}\tan\alpha - \frac{0.01181}{\cos^2\alpha}\left(\frac{42.34}{\tan\alpha}\right)^2 - 11 = 0$$

$$\Rightarrow 42.34 - \frac{21.17}{\sin^2\alpha} - 11 = 0$$

$$\Rightarrow 31.34 - \frac{21.17}{\sin^2\alpha} = 0$$

$$\Rightarrow \sin^2\alpha = \frac{21.17}{31.34}$$

$$\Rightarrow \sin^2\alpha = 0.6755$$

$$\Rightarrow \sin\alpha = \sqrt{0.6755} = 0.8218$$

$$\Rightarrow \alpha = \sin^{-1}(0.8218)$$

$$\Rightarrow \alpha = 0.9646 \text{ rad} = 55^0 27'$$

Put the value of α in the expression for x, we obtain as

$$x = \frac{42.34}{\tan \alpha} = \frac{42.34}{\tan \left(55^0 27' \right)}$$

$$\Rightarrow x = \frac{42.34}{1.44} = 29.41 \text{ m}$$

4. UTILITY ON TURBOMACHINERY

Case 1. APPLICABILITY ON HYDRO POWER PLANT

In a hydroelectric power plant, water flows at the rate of $120 \text{ m}^3/\text{s}$ from an elevation of 200 m to a turbine, where electric power is generated as shown in figure. The total irreversible head loss in the piping system from point 1 to point 2 (excluded the turbine unit) is 40 m. If overall efficiency of the turbine generator is 80 %, calculate the electric power output.

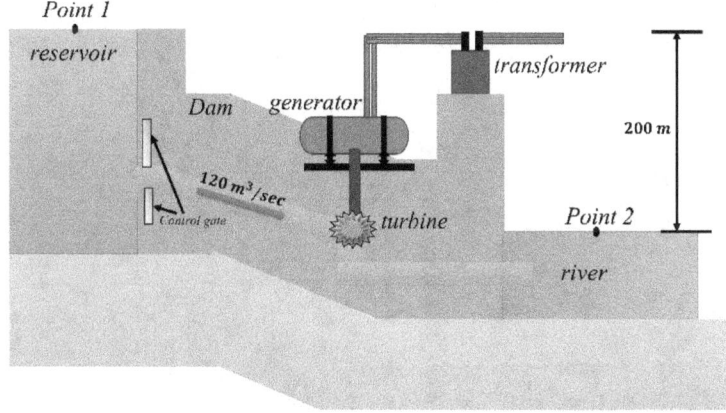

Discussion: In this case,

$Q =$ discharge rate of flow $= 120 \text{ m}^3/\text{s}$

$z_1 =$ elevation of water level from point $1 = 200$ m

$h_l =$ irreversible head loss in piping system $= 40$ m

$\varepsilon_{gen} =$ overall efficiency of turbine generator $= 80$ %

Let $h_{turbine}$ = head loss in turbine's shaft

ρ = density of water = 1000 kg/m^3

The mass flow rate of water through turbine is

$$\dot{M} = \rho Q = 1000 \times 120 = 1.2 \times 10^5 \text{ kg/s}$$

Consider the point 1 is at the free surface of water reservoir and point 2 at the water surface close to turbine generator. Now apply Bernoulli's equation along the streamline of water flow through point 1 to the point 2 is

$$\frac{P_1}{\rho g} + \frac{v_1^2}{2g} + z_1 = \frac{P_2}{\rho g} + \frac{v_2^2}{2g} + z_2 + h_{turbine} + h_l \quad \text{------(1)}$$

Since both the point 1 and point 2 are on the free surface of water, so $P_1 = P_2 = P_{atm}$ Thus, equation (1) reduced to

$$\frac{P_{atm}}{\rho g} + \frac{v_1^2}{2g} + z_1 = \frac{P_{atm}}{\rho g} + \frac{v_2^2}{2g} + z_2 + h_{turbine} + h_l$$

$$\Rightarrow \frac{v_1^2}{2g} + z_1 = \frac{v_2^2}{2g} + z_2 + h_{turbine} + h_l \quad \text{------(2)}$$

Since velocity of water flow through streamline assumed to very small such that $v_1 = v_2 = 0$. Thus, equation (2) reduced to

$$0 + z_1 = 0 + z_2 + h_{turbine} + h_l$$
$$\Rightarrow z_1 = z_2 + h_{turbine} + h_l \quad \text{------(3)}$$

Since point 2 taken at reference level, thus we have $z_2 = 0$ and equation (3) reduced to

$$z_1 = h_{turbine} - h_l$$

Put the value of z_1 and h_l in the above expression, we get

$$h_{turbine} = z_1 - h_l = 200 - 40 = 160 \text{ m}$$

Next, the shaft power output through turbine's shaft is

$$W_{turbine} = \dot{M}\, gh_{turbine} = 1.2 \times 10^5 \times 9.81 \times 160$$
$$\Rightarrow W_{turbine} = 188352000 \text{ kg m}^2/\text{s}^3 = 188352 \text{ kW}$$

The electric power generated by the actual unit is

$$W = W_{turbine} \times \varepsilon_{gen} = 188352 \times 0.8$$
$$\Rightarrow W = 150681.6 \text{ kW} = 150.68 \text{ MW}$$

Case 2. APPLICABILTY ON HYDRAULIC TURBINE

A hydraulic turbine has 50 m of head available at a flow rate of $1.30 \text{ m}^3/\text{s}$, as shown in figure and its overall turbine-generator efficiency is 80 % . The elevation of water level from the reference line is 150 m. Determine the electric power output of this turbine generator.

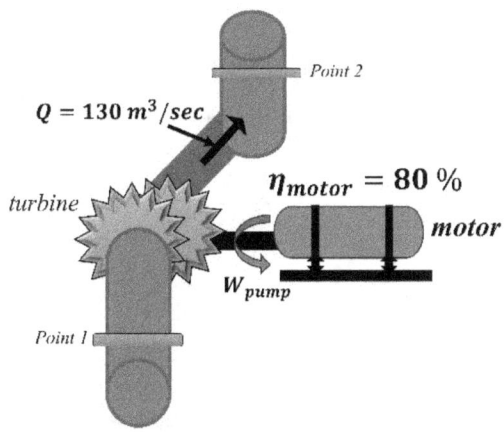

Discussion: In this case,

$Q =$ discharge rate of flow $= 130 \text{ m}^3/\text{s}$

Let z_1 = elevation of water level from point 1 $= 150$ m

h_l = irreversible head loss in piping system $= 50$ m

ε_{gen} = overall efficiency of turbine generator $= 80\ \%$

Let $h_{turbine}$ = head loss in turbine's shaft

ρ = density of water $= 1000$ kg/m^3

The mass flow rate of water through turbine is

$$\dot{M} = \rho Q = 1000 \times 130 = 1.3 \times 10^5 \text{ kg/s}$$

Consider the point 1 is at the free surface of water and point 2 at the water surface close to turbine generator. Now apply Bernoulli's equation along the streamline of water flow through point 1 to the point 2 is

$$\frac{P_1}{\rho g} + \frac{v_1^2}{2g} + z_1 = \frac{P_2}{\rho g} + \frac{v_2^2}{2g} + z_2 + h_{turbine} + h_l \quad \text{------(1)}$$

Since both the point 1 and point 2 are on the free surface of water, so $P_1 = P_2 = P_{atm}$ Thus, equation (1) reduced to

$$\frac{P_{atm}}{\rho g} + \frac{v_1^2}{2g} + z_1 = \frac{P_{atm}}{\rho g} + \frac{v_2^2}{2g} + z_2 + h_{turbine} + h_l$$

$$\Rightarrow \frac{v_1^2}{2g} + z_1 = \frac{v_2^2}{2g} + z_2 + h_{turbine} + h_l \quad \text{------(2)}$$

Since velocity of water flow through streamline assumed to very small such that $v_1 = v_2 = 0$. Thus, equation (2) reduced to

$$0 + z_1 = 0 + z_2 + h_{turbine} + h_l$$
$$\Rightarrow z_1 = z_2 + h_{turbine} + h_l \quad \text{------(3)}$$

Since point 2 taken at reference level, thus we have $z_2 = 0$ and equation (3) reduced to

$$z_1 = h_{turbine} - h_l$$

Put the value of z_1 and h_l in the above expression, we get

$$h_{turbine} = z_1 - h_l = 150 - 50 = 100 \text{ m}$$

Next, the shaft power output through turbine's shaft is

$$W_{turbine} = \dot{M} gh_{turbine} = 1.3 \times 10^5 \times 9.81 \times 100$$
$$\Rightarrow W_{turbine} = 127530000 \text{ kg m}^2 / \text{s}^3 = 127530 \text{ kW}$$

The electric power generated by the actual unit is

$$W = W_{turbine} \times \varepsilon_{gen} = 127530 \times 0.8$$
$$\Rightarrow W = 102024 \text{ kW} = 102.02 \text{ MW}$$

Case 3. APPLICABILITY ON COMPUTER COOLER

A fan chosen to cool a computer case whose dimension are $15 \times 30 \times 30$ cm^3. Half of the case is filled with components and other half is filled with air. A 5 cm diameter hole is available at the back of the case as shown in figure for the installation of the fan that is to replace the air in the void space of the case once every second. Small low-power fan-motor combined units are available in the market and its efficiency is 30 %. Determine the wattage of the fan-motor unit.

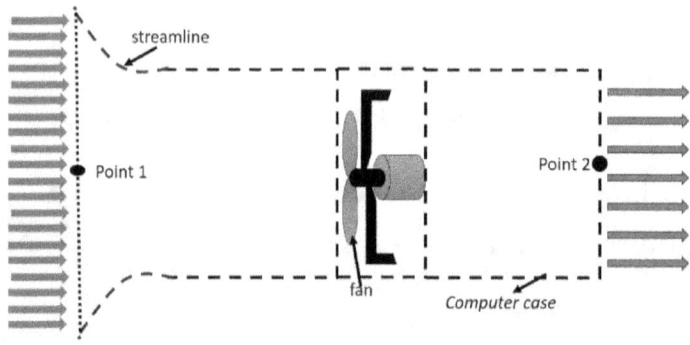

Discussion: In this case,

V = volume of the computer case $= 15 \times 30 \times 30$ cm^3

φ = void fraction in the case $= \dfrac{1/2V}{V} = \dfrac{1}{2}$

d = diameter of open hole $= 5$ cm $= 0.05$ m

ρ = density of air $= 1.2$ kg/m^3

α = kinetic energy correction factor $= 1.10$

η_{motor} = efficiency of fen motor $= 30\% = 0.3$

Next, fraction of air occupied in the computer case is

$$\delta V_{air} = V \times \varphi = 15 \times 30 \times 30 \times \frac{1}{2} = 6750 \text{ cm}^3$$

Thus, rate of change in volume of air through the case is

$$\dot{V} = \frac{\delta V_{air}}{\delta t} = \frac{6750}{1} = 6.75 \times 10^{-3} \text{ m}^3/\text{sec}$$

The rate of mass flow of air through the case is

$$\dot{M} = \rho \dot{V} = 1.2 \times 6.75 \times 10^{-3} = 0.0081 \text{ kg/sec}$$

Next, cross-sectional area of hole is

$$a = \frac{\pi}{4}d^2 = \frac{\pi}{4}(0.05)^2 = 1.96 \times 10^{-3} \text{ m}^2$$

The velocity of air pass through outlet is

$$v = \frac{\dot{V}}{a} \equiv \frac{Q}{A} = \frac{6.75 \times 10^{-3}}{1.96 \times 10^{-3}} = 3.44 \text{ m/sec}$$

Consider the point 1 is at inlet which is at a certain distance from the fan and point 2 is at outlet of fan in the case. In this case, both the point is open to atmosphere so that we take, $P_1 = P_2 = P_{atm}$. Also, $v_1 : 0$ as the point is at the certain distance from the fan-engine. In order to calculate the power output of motor, we assumed that there is negligible frictional loss and both the point 1 and point 2 are at same level such that $z_1 = z_2$.

On apply the Bernoulli's equation along the streamline of air flow through the point 1 to point 2, we obtain as

$$W_{shaft} + \frac{P_1}{\rho_1} + \frac{v_1^2}{2} + gz_1 = \frac{P_2}{\rho_1} + \frac{v_2^2}{2} + gz_2 + E_{loss}$$

$$\Rightarrow W_{fan} - W_{motor} + \frac{P_1}{\rho_1} + \frac{v_1^2}{2} + gz_1 = \frac{P_2}{\rho_1} + \frac{v_2^2}{2} + gz_2 + E_{loss}$$

$$\Rightarrow W_{fan} + \frac{P_1}{\rho_1} + \frac{v_1^2}{2} + gz_1 = W_{motor} + \frac{P_2}{\rho_1} + \frac{v_2^2}{2} + gz_2 + E_{loss}$$

On multiply the above equation with mass flow rate, we get the energy equation which is defined as

$$W_{fan} + \dot{M}\left(\frac{P_1}{\rho} + \frac{v_1^2}{2} + gz_1\right) = W_{motor} + \dot{M}\left(\frac{P_2}{\rho} + \frac{v_2^2}{2} + gz_2\right) + E_{loss} \quad \text{-----(1)}$$

Since $P_1 = P_2 = P_{atm} \,|\, v_1 : \; 0 \,|\, z_1 = z_2$, the equation (1) reduced to

$$W_{fan} = \dot{M}\left(\frac{v_2^2}{2}\right) + E_{loss} \quad \text{-----(2)}$$

Due to the variation of velocity of air flows through motor, we introduce the correction factor, α to the kinetic energy term $\dfrac{v_2^2}{2}$ in the above energy equation. Thus, equation (2) can be written as

$$W_f = W_{fan} - E_{loss} = \dot{M}\,\alpha\,\frac{v_2^2}{2}$$

$$\Rightarrow W_f = 0.0081 \times 1.1 \times \frac{(3.44)^2}{2} = 0.0527 \text{ W}$$

Finally, the required electric power required to the fan-motor is

$$P = \frac{W_f}{\eta_{motor}} = \frac{0.0527}{0.3} = 0.175 \text{ W}$$

Case 4. APPLICABILTY ON ROOM VENTILATION

A fan is to be chosen to ventilate a bathroom whose dimension are $2 \times 3 \times 3$ m^3 as shown in figure. The air velocity is not to exceed 7 m/s to minimize vibration and noise. The combined efficiency of the fan-motor unit to be used can be taken as 50 %. If the fan is to replace the entire volume of air in 15 min, determine the wattage of the fan-motor unit to be purchased provided the density of air is 1.25 kg/m^3.

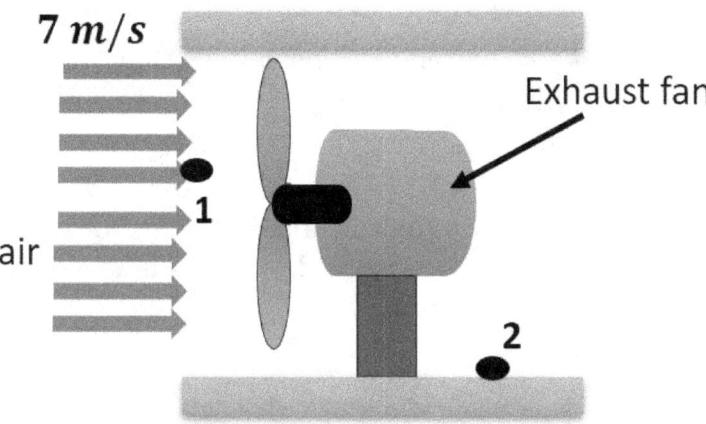

Discussion: In this case,

V = volume of the room $= 2 \times 3 \times 3 \text{ m}^3$

v = velocity of air flow $= 7$ m/sec

ρ = density of air $= 1.2 \text{ kg/m}^3$

α = kinetic energy correction factor $= 1.10$

η_{motor} = efficiency of fen motor $= 50 \% = 0.5$

Thus, rate of change in volume of air through the room is

$$\dot{V} = \frac{V_{air}}{t} = \frac{18}{1} = 18 \text{ m}^3 / \text{sec}$$

The rate of mass flow of air through the room is

$$\dot{M} = \rho \dot{V} = 1.2 \times 18 = 21.6 \text{ kg/sec}$$

Consider the point 1 is at inlet which is at a certain distance from the fan and point 2 is at outlet of fan in the bathroom. In this case, both the point are open to atmosphere so that we take, $P_1 = P_2 = P_{atm}$. Also, $v_1 : 0$ as the point is at the certain distance from the fan-engine. In order to calculate the power output of motor, we assumed that there is negligible frictional loss and both the point 1 and point 2 are at same level such that $z_1 = z_2$.

On apply the Bernoulli's equation along the streamline of air flow through the point 1 to point 2, we obtain as

$$W_{shaft} + \frac{P_1}{\rho_1} + \frac{v_1^2}{2} + gz_1 = \frac{P_2}{\rho_1} + \frac{v_2^2}{2} + gz_2 + E_{loss}$$

$$\Rightarrow W_{fan} - W_{motor} + \frac{P_1}{\rho_1} + \frac{v_1^2}{2} + gz_1 = \frac{P_2}{\rho_1} + \frac{v_2^2}{2} + gz_2 + E_{loss}$$

$$\Rightarrow W_{fan} + \frac{P_1}{\rho_1} + \frac{v_1^2}{2} + gz_1 = W_{motor} + \frac{P_2}{\rho_1} + \frac{v_2^2}{2} + gz_2 + E_{loss}$$

On multiply the above equation with mass flow rate, we get the energy equation which is defined as

$$W_{fan} + \dot{M}\left(\frac{P_1}{\rho} + \frac{v_1^2}{2} + gz_1\right) = W_{motor} + \dot{M}\left(\frac{P_2}{\rho} + \frac{v_2^2}{2} + gz_2\right) + E_{loss} \quad \text{-----(1)}$$

Since $P_1 = P_2 = P_{atm} \mid v_1 : 0 \mid z_1 = z_2$, the equation (1) reduced to

$$W_{fan} = \dot{M}\left(\frac{v_2^2}{2}\right) + E_{loss} \quad \text{-----(2)}$$

Due to the variation of velocity of air flows through motor, we introduce the correction factor, α to the kinetic energy term $\frac{v_2^2}{2}$ in the above energy equation. Thus, equation (2) can be written as

$$W_f = W_{fan} - E_{loss} = \dot{M}\,\alpha\,\frac{v_2^2}{2}$$

$$\Rightarrow W_f = 21.6 \times 1.1 \times \frac{(7)^2}{2} = 582.12 \text{ W}$$

Finally, the required electric power required to the fan-motor is

$$P = \frac{W_f}{\eta_{motor}} = \frac{582.12}{0.5} = 1164.24 \text{ W}$$

Case 5. APPLICABILTY ON SUBMERSIBLE PUMP

A submersible pump with a shaft power of 5 kW and an efficiency of 80 % is used to pump water from a lake to a pool through a constant diameter pipe as shown in figure. The free surface of the pool is 30 m above the free surface of the lake. If the irreversible head loss in the piping system is 4 m. Determine the discharge rate of water.

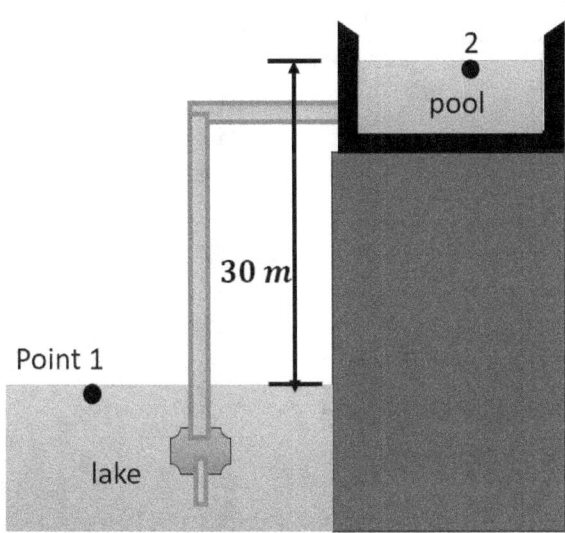

Discussion: In this case,

W_{shaft} = power output of shaft = 5 kW

η_{pump} = efficiency of pump = 80 % = 0.8

133

z_2 = elevation of point 2 = 30 m

h_l = head loss in piping system = 4 m

Next, the power supply to water is

$$W_{pump} = W_{shaft}\eta_{pump} = 5 \times 0.8 = 4 \text{ kW}$$

Consider the point 1 is at the free surface of the lake, which assumed as the reference line such that $z_1 = 0$. Also, the point 2 is at the centre of pool. Since, both the points are open to atmosphere so that $P_1 = P_2 = P_{atm}$ and assume that velocities of water flow is very small such that $v_1 = v_2 : 0$.

On apply the Bernoulli's equation along the streamline of air flow through the point 1 to point 2, we obtain as

$$W_{shaft} + \frac{P_1}{\rho_1} + \frac{v_1^2}{2} + gz_1 = \frac{P_2}{\rho_1} + \frac{v_2^2}{2} + gz_2 + E_{loss}$$

$$\Rightarrow W_{pump} - W_{turbine} + \frac{P_1}{\rho_1} + \frac{v_1^2}{2} + gz_1 = \frac{P_2}{\rho_1} + \frac{v_2^2}{2} + gz_2 + E_{loss}$$

$$\Rightarrow W_{pump} + \frac{P_1}{\rho_1} + \frac{v_1^2}{2} + gz_1 = W_{turbine} + \frac{P_2}{\rho_1} + \frac{v_2^2}{2} + gz_2 + E_{loss}$$

On multiply the above equation with mass flow rate, we get the energy equation which is defined as

$$W_{pump} + \dot{M}\left(\frac{P_1}{\rho} + \frac{v_1^2}{2} + gz_1\right) = W_{turbine} + \dot{M}\left(\frac{P_2}{\rho} + \frac{v_2^2}{2} + gz_2\right) + E_{loss} \quad \text{-----(1)}$$

Since $P_1 = P_2 = P_{atm} \mid v_1 = v_2 \equiv 0 \mid z_1 = 0$, the equation (1) reduced to

$$W_{pump} = \dot{M}\, gz_2 + E_{loss} \quad \text{------(2)}$$

Since $E_{loss} = \dot{M} gh_l$, The equation (2) reduced to

$$W_{pump} = \dot{M} gz_2 + \dot{M} gh_l$$

$$\Rightarrow W_{pump} = \dot{M} g(z_2 + h_l)$$

$$\Rightarrow \dot{M} = \frac{W_{pump}}{g(z_2 + h_l)}$$

$$\Rightarrow \dot{M} = \frac{4}{9.81 \times (30 + 4)} \; ; \; 12.1 \text{ kg/sec}$$

Rate of discharge of water is

$$Q = \frac{\dot{M}}{\rho} = \frac{12.1}{1000} = 12.1 \times 10^{-3} \text{ m}^3 / \text{sec} = 12.1 \text{ Lt/sec}$$

Case 6. APPLICABILTY ON PUMPING SYSTEM

Underground water is to be pumped by a 78 % efficient 5 kW submerged pump to a pool whose free surface is 30 m above the underground water level. The diameter of the pipe is 7 cm on the intake side and 5 cm on the discharge side. Determine the maximum flow rate of water.

Discussion: In this case,

W_{shaft} = power output of shaft = 5 kW

η_{pump} = efficiency of pump = 78 % = 0.78

z_2 = elevation of point 2 = 30 m

Next, the power supply to water is

$$W_{pump} = W_{shaft}\eta_{pump} = 5 \times 0.78 = 3.6 \text{ kW}$$

Consider the point 1 is at the free surface of the lake, which assumed as the reference line such that $z_1 = 0$. Also, the point 2 is at the centre of pool. Since, both the points are open to atmosphere so that $P_1 = P_2 = P_{atm}$ and assume that velocities of water flow is very small such that $v_1 = v_2 : 0$.

On apply the Bernoulli's equation along the streamline of air flow through the point 1 to point 2, we obtain as

$$W_{shaft} + \frac{P_1}{\rho_1} + \frac{v_1^2}{2} + gz_1 = \frac{P_2}{\rho_1} + \frac{v_2^2}{2} + gz_2 + E_{loss}$$

$$\Rightarrow W_{pump} - W_{turbine} + \frac{P_1}{\rho_1} + \frac{v_1^2}{2} + gz_1 = \frac{P_2}{\rho_1} + \frac{v_2^2}{2} + gz_2 + E_{loss}$$

$$\Rightarrow W_{pump} + \frac{P_1}{\rho_1} + \frac{v_1^2}{2} + gz_1 = W_{turbine} + \frac{P_2}{\rho_1} + \frac{v_2^2}{2} + gz_2 + E_{loss}$$

On multiply the above equation with mass flow rate, we get the energy equation which is defined as

$$W_{pump} + \dot{M}\left(\frac{P_1}{\rho} + \frac{v_1^2}{2} + gz_1\right) = W_{turbine} + \dot{M}\left(\frac{P_2}{\rho} + \frac{v_2^2}{2} + gz_2\right) + E_{loss} \quad \text{-----(1)}$$

Since $P_1 = P_2 = P_{atm} \mid v_1 = v_2 \equiv 0 \mid z_1 = 0,$ the equation (1) reduced to

$$W_{pump} = \dot{M} gz_2 + E_{loss} \quad \text{------(2)}$$

Since $E_{loss} = \dot{M} gh_l,$ The equation (2) reduced to

$$W_{pump} = \dot{M}\, gz_2 + \dot{M}\, gh_l$$

$$\Rightarrow W_{pump} = \dot{M}\, g\left(z_2 + h_l\right)$$

$$\Rightarrow \dot{M} = \frac{W_{pump}}{g\left(z_2 + h_l\right)}$$

$$\Rightarrow \dot{M} = \frac{3.6}{9.81 \times (30 + 0)}\; ;\; 12.2 \text{ kg/sec}$$

Rate of discharge of water is

$$Q = \frac{\dot{M}}{\rho} = \frac{12.2}{1000} = 12.2 \times 10^{-3} \text{ m}^3/\text{sec} = 12.2 \text{ Lt/sec}$$

Case 7: APPLICIABILITY ON PIPE BENDINGS

PROBLEM 1. A 45° reducing bend is connected in a pipe line as shown in figure, the diameters at the inlet and outlet of the bend being 40 cm and 20 cm respectively. Find the pressure at outlet of pipe bend if the intensity of pressure at inlet of pipe bend is 21.58 N/cm^2. The rate of flow of water is 500 Lt/sec.

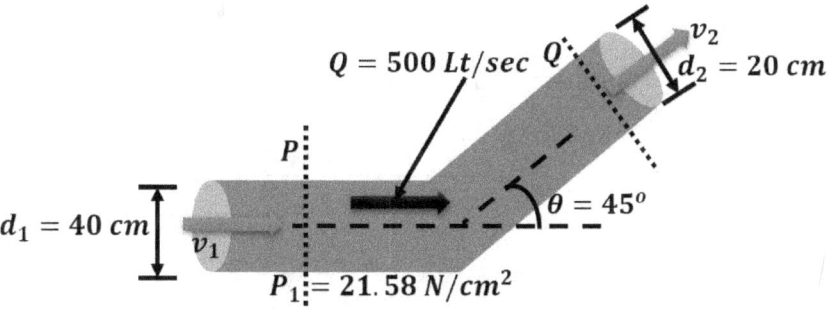

Solution: In this case,

D = diameter of inlet of pipe = 40 cm = 0.4 m

A = cross-sectional area of inlet $= \dfrac{\pi}{4}D^2 = \dfrac{\pi}{4}(0.4)^2 = 0.1256 \text{ m}^2$

d = diameter of outlet of the pipe $= 20 \text{ cm} = 0.2 \text{ m}$

a = cross-sectional area of outlet $= \dfrac{\pi}{4}d^2 = \dfrac{\pi}{4}(0.2)^2 = 0.0314 \text{ m}^2$

Q = discharge rate of flow of water $= 500 \text{ Lt/sec} = 0.5 \text{ m}^3 / \text{sec}$

v_1 = velocity of flow of water at section $\text{P} = \dfrac{Q}{A} = \dfrac{0.5}{0.1256} = 4 \text{ m/s}$

v_2 = velocity of flow of water at section Q

$= \dfrac{Q}{a} = \dfrac{0.5}{0.0314} = 16 \text{ m/s}$

P_1 = pressure at inlet $= 21.58 \text{ N/cm}^2 = 21.58 \times 10^4 \text{ N/m}^2$

Next applying Bernoulli's equation at section P and Q, we get

$$\frac{P_1}{\rho g} + \frac{v_1^2}{2g} + z_1 = \frac{P_2}{\rho g} + \frac{v_2^2}{2g} + z_2$$

In this case, $\quad z_1 = z_2$

The Bernoulli's equation reduced to

$$\frac{P_1}{\rho g} + \frac{v_1^2}{2g} = \frac{P_2}{\rho g} + \frac{v_2^2}{2g}$$

$$\Rightarrow \frac{P_1}{\rho} + \frac{v_1^2}{2} = \frac{P_2}{\rho} + \frac{v_2^2}{2}$$

$$\Rightarrow \frac{P_2}{\rho} = \frac{P_1}{\rho} + \frac{v_1^2}{2} - \frac{v_2^2}{2}$$

$$\Rightarrow \frac{P_2}{\rho} = \frac{21.58 \times 10^4}{10^3} + \frac{1}{2}(4)^2 - \frac{1}{2}(16)^2$$

$$\Rightarrow \frac{P_2}{\rho} = 215.8 + 8 - 128 = 95.8$$

$$\Rightarrow P_2 = 95.8 \times 1000 = 9.58 \times 10^4 \text{ N/m}^2$$

PROBLEM 2. A 30 cm diameter pipe carries water under a head of 15 m with a velocity of 4 m/s. If the x-axis of the pipe turns through 45°, as shown in figure. Determine the magnitude and direction of the resultant force at the bend.

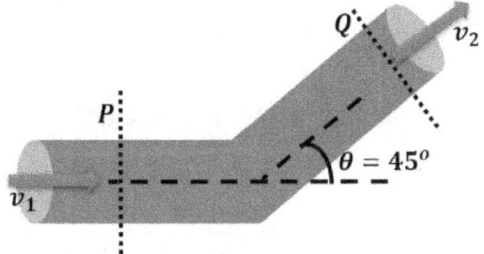

Solution: In this case,

$d = d_{inlet} = d_{oulet} = $ diameter of pipe bend 30 cm = 0.3 m

$\therefore$ cross-sectional area, $A = \frac{\pi}{4}d^2 = \frac{\pi}{4}(0.3)^2 = 0.07068 \text{ m}^2$

$v = v_1 = v_2 = $ velocity of flow $= 4$ m/s

$Q = $ discharge $= 0.07068 \times 4 = 0.28272 \text{ m}^3/\text{s}$

$\frac{P}{\rho g} = $ pressure head $= 15$ m of water

$\Rightarrow P = P_1 = P_2 = 15 \times \rho g = 15 \times 1000 \times 9.81 = 147150 \text{ N/m}^2$

Next, $v_{1x} = $ initial velocity along x-axis $= 4$ m/s

v_{2x} = final velocity in x-axis

$= v_2 \cos 45^o = 4 \times \dfrac{1}{\sqrt{2}} = 4 \times 0.7071 = 2.8284$ m/s

v_{1y} = initial velocity along y-axis $= 0$

v_{2y} = final velocity along y-axis

$= v_{2y} \sin 45^o = 4 \times 0.7071 = 2.8284$ m/s

$\left(P_1 A\right)_x$ = pressure force exerted at section P along x-axis

$= 147150 \times 0.07068 = 10400.562$

$\left(P_1 A\right)_y$ = pressure force exerted at section P along y-axis $= 0$

$\left(P_2 A\right)_x$ = pressure force exerted at section Q along x-axis

$= -P_2 A \cos 45^o = -147150 \times 0.07068 \times 0.7071 = 7354$

$\left(P_2 A\right)_y$ = pressure force exerted at section Q along y-axis

$= -P_2 A \sin 45^o = -147150 \times 0.07068 \times 0.7071 = 7354$

Force exerted along x-axis $F_x = \rho Q \left[v_{1x} - v_{2x}\right] + \left(P_1 A\right)_x + \left(P_2 A\right)_x$

$$F_x = 1000 \times 0.28272 \left[4 - 2.8284\right] + 10400.562 + 7354$$
$$\Rightarrow F_x = 331.234 + 10400.562 + 7354$$
$$\Rightarrow F_x = 18085.796$$

Force exerted along y-axis $F_y = \rho Q \left[v_{1y} - v_{2y}\right] + \left(P_1 A\right)_y + \left(P_2 A\right)_y$

$$F_y = 1000 \times 0.28272 \left[0 - 2.8284\right] + 0 + 7354$$
$$\Rightarrow F_y = 6554.35$$

$\therefore$ Resultant force, $F = \sqrt{F_x^2 + F_y^2}$

$$\Rightarrow F = \sqrt{\left(18085.796\right)^2 + \left(6554.35\right)^2}$$

$$\Rightarrow F = \sqrt{327096016.954 + 42959503.9225}$$

$$\Rightarrow F = \sqrt{370055520.876}$$

$$\Rightarrow F = 19237$$

The angle made by resultant force with $x-$ axis

$$= \tan \theta = \frac{F_y}{F_x} = \frac{6554.35}{18085.796}$$

$$\Rightarrow \theta = \tan^{-1}\left(0.3624\right)$$

$$\Rightarrow \theta = 0.3477 \text{ rad} = 18°22'$$

PROBLEM 3. The discharge of water through a pipe of diameter 40 cm is 400 Lt/sec. If the pipe is bend by $135°$, as shown in figure. Determine the magnitude and direction of the resultant force on the bend. The pressure of flowing water is 29.43 N/cm^2.

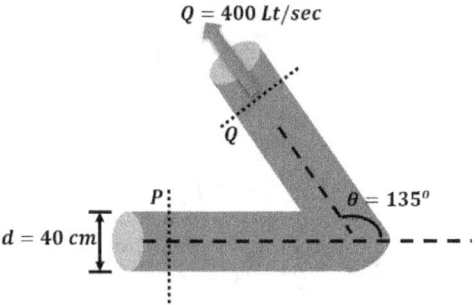

$Q = 400\ Lt/sec$

P

$\theta = 135°$

$d = 40\ cm$

Solution: In this case,

$d = d_{inlet} = d_{oulet} =$ diameter of pipe bend 40 cm $= 0.4$ m

$\therefore$ cross-sectional area, $A = \frac{\pi}{4} d^2 = \frac{\pi}{4}\left(0.4\right)^2 = 0.1256$ m^2

$Q =$ discharge $= 400$ Lt/sec $= 0.4$ m$^3/$s

$v = v_1 = v_2 =$ velocity of flow

$$\Rightarrow v = \frac{Q}{A} = \frac{0.4}{0.1256} = 3.18 \text{ m/s}$$

$P = P_1 = P_2 =$ pressure at the section P and Q

$= 29.43 \text{ N/cm}^2 = 29.43 \times 10^4 \text{ N/m}^2$

Next, $v_{1x} =$ initial velocity along x-axis $= 3.18$ m/s

$v_{2x} =$ final velocity in x-axis

$= -v_2 \cos 45^\circ = -3.18 \times \dfrac{1}{\sqrt{2}} = -3.18 \times 0.7071 = -2.25$ m/s

$v_{1y} =$ initial velocity along y-axis $= 0$

$v_{2y} =$ final velocity along y-axis

$= -v_2 \sin 45^\circ = -3.18 \times 0.7071 = -2.25$ m/s

$\left(P_1 A \right)_x =$ pressure force exerted at section P along x-axis

$= 29.43 \times 10^4 \times 0.1256 = 36964.08$

$\left(P_1 A \right)_y =$ pressure force exerted at section P along y-axis $= 0$

$\left(P_2 A \right)_x =$ pressure force exerted at section Q along x-axis

$= P_2 A \cos 45^\circ = 29.43 \times 10^4 \times 0.1256 \times 0.7071 = 26137.30$

$\left(P_2 A \right)_y =$ pressure force exerted at section Q along y-axis

$= -P_2 A \sin 45^\circ = -29.43 \times 10^4 \times 0.1256 \times 0.7071 = -26137.30$

Force exerted along x-axis $F_x = \rho Q \left[v_{1x} - v_{2x} \right] + \left(P_1 A \right)_x + \left(P_2 A \right)_x$

$$F_x = 1000 \times 0.4 \left[3.18 + 2.25 \right] + 36964.08 + 26137.30$$
$$\Rightarrow F_x = 2172 + 36964.08 + 26137.30$$
$$\Rightarrow F_x = 65273.38$$

Force exerted along y-axis $F_y = \rho Q \left[v_{1y} - v_{2y} \right] + \left(P_1 A \right)_y + \left(P_2 A \right)_y$

$$F_y = 1000 \times 0.4 \left[0 - 2.25 \right] + 0 - 26137.30$$
$$\Rightarrow F_y = -900 + 0 - 26137.30$$
$$\Rightarrow F_y = -27037.3$$

∴   Resultant force, $F = \sqrt{F_x^2 + F_y^2}$

$$\Rightarrow F = \sqrt{\left(65273.38 \right)^2 + \left(-27037.30 \right)^2}$$

$$\Rightarrow F = \sqrt{4260614136.62 + 731015591.29}$$
$$\Rightarrow F = \sqrt{4991629727.91}$$
$$\Rightarrow F = 70651$$

The angle made by resultant force with $x-$ axis

$$= \tan \theta = \frac{F_y}{F_x} = \frac{27037.30}{65273.38}$$
$$\Rightarrow \theta = \tan^{-1} \left(0.4142 \right)$$
$$\Rightarrow \theta = 0.3927 \text{ rad} = 22^\circ 5'$$

PROBLEM 4. A pipe of 20 cm diameter carries $0.20 \text{ m}^3 / \sec$ of water has a right angled bend in a horizontal plane as shown in figure. Determine the resultant force exerted on the bend if the pressure at inlet and outlet of the pipe bend are 22.563 N/cm^2 and 21.582 N/cm^2 respectively.

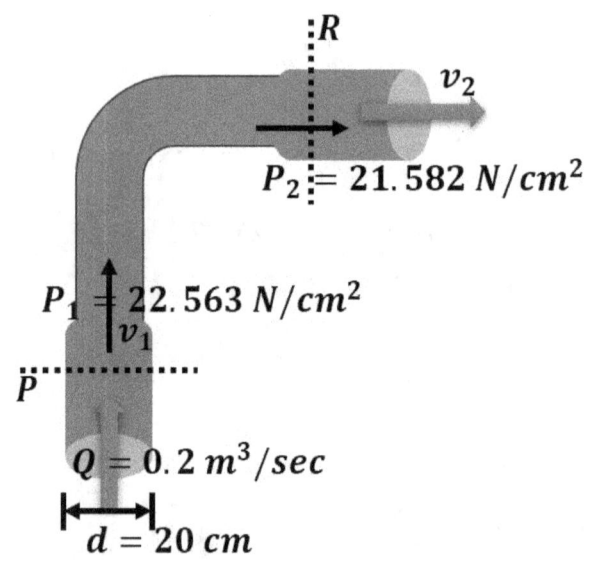

$$R$$

$$v_2$$

$$P_2 = 21.582\ N/cm^2$$

$$P_1 = 22.563\ N/cm^2$$

$$v_1$$

$$P$$

$$Q = 0.2\ m^3/sec$$

$$d = 20\ cm$$

Solution: In this case,

d = diameter of pipe = 20 cm = 0.2 m

∴ cross-sectional area, $A = A_P = A_Q = \dfrac{\pi}{4}(0.2)^2$

$$\Rightarrow A = 0.0314\ m^2$$

Q = discharge = 0.20 m³ / sec

∴ velocity $v = v_1 = v_2 = \dfrac{Q}{A} = \dfrac{0.2}{0.0314} = 6.36$ m/s

$P_1 = 22.563\ N/cm^2 = 22.563 \times 10^4\ N/m^2$

$P_2 = 21.582\ N/cm^2 = 21.582 \times 10^4\ N/m^2$

Force on bend along $x-$ axis, $F_x = \rho Q[v_{1x} - v_{2x}] + (P_1 A_1)_x + (P_2 A_2)_x$

Where ρ = density of water = 1000 kg/m³

$v_{1x} = v =$ velocity of flow at section P along x-axis $= 6.36$ m/s

$v_{2x} =$ velocity of flow at section Q along x-axis $= 0$

$P_1 A_1 = P_1 A = 22.563 \times 10^4 \times 0.0314 = 0.71 \times 10^4$

$P_2 A_2 = P_2 A = 0$

$\therefore \quad F_x = 1000 \times 0.2[6.36 - 0] + 7100 + 0 \Rightarrow F_x = 8372$

Force on bend along $y-$ axis,

$F_y = \rho Q \left[v_{1y} - v_{2y} \right] + \left(P_1 A_1 \right)_y + \left(P_2 A_2 \right)_y$

Where $\rho =$ density of water $= 1000$ kg/m^3

$v_{1y} =$ velocity of flow at section P along ẏ-axis $= 0$

$v_{2y} =$ velocity of flow at section Q along y-axis $= v = 6.36$ m/s

$\left(P_1 A_1 \right)_y = \left(P_1 A \right)_y =$ force exerted by flow at the section P $= 0$

$\left(P_2 A_2 \right)_y = 22.582 \times 10^4 \times 0.0314 = 0.71 \times 10^4$

$\therefore \quad F_y = 1000 \times 0.2[0 - 6.36] + 0 + 0.71 \times 10^4 \Rightarrow F_y = 5828$

$\therefore \quad$ Resultant force, $F = \sqrt{F_x^2 + F_y^2} = \sqrt{(8372)^2 + (5828)^2}$

$\Rightarrow F = \sqrt{70090384 + 33965584}$

$\Rightarrow F = \sqrt{104055968}$

$\Rightarrow F = 10200.78$